WATERSHED MANAGEMENT

PROCEEDINGS OF THE THIRD INTERNATIONAL CONFERENCE

December 11–14, 2001
National Taiwan University
Taipei, Taiwan

SPONSORED BY
Environmental and Water Resources Institute (EWRI) of ASCE
Products Development

EDITED BY
Michael A. Ports
David W. Renetzky

ASCE American Society of Civil Engineers

EWRI

Published by the American Society of Civil Engineers

Library of Congress Cataloging-in-Publication Data

International Conference on Watershed Management (3rd : 2001 : Taipei, Taiwan)
Watershed management : proceedings of the third international conference / edited by Michael A. Ports and David W. Renetzky.
p. cm.
"Third International Conference on Watershed Management, held Dec. 11-14, 2001 at National Taiwan University in Taipei, Taiwan"--Introd.
Also issued under title : Di san jie ji shui qu guan li guo ji yan tao hui lun wen ji.
Includes bibliographical references and index.
Includes some papers translated from Chinese.
ISBN 0-7844-0706-1
1. Watershed management--Congresses. 2. Water quality management--Congresses. 3. Water-supply--Management--Congresses. I. Ports, Michael A. II. Renetzky, David W. III. Title.

TC409.I565 2003
627--dc22 2003062665

American Society of Civil Engineers
1801 Alexander Bell Drive
Reston, Virginia, 20191-4400

www.pubs.asce.org

Library of Congress Catalog Card No: 2003062665
ISBN 0-7844-0706-1
Manufactured in the United States of America.

Introduction

The Third International Conference on Watershed Management was held December 11-14, 2001 at National Taiwan University in Taipei, Taiwan. The conference was successful in meeting its objective of gathering water management experts from the United States, Japan, and Taiwan to discuss water supply and water pollution issues in a sustainable context. The issues are of critical importance not only to Taiwan, but countries throughout the Pacific Rim. Over twenty-five papers were presented by invited speakers on subjects such as new technologies for erosion and sediment control, non-point source and urban stormwater pollution reduction, and hazard mitigation. Facilitated round table discussions focused on sustainable watershed management strategies, debris flow, and river restoration. The conference concluded with a technical tour of Taiwan's Feitsui Reservoir and the surrounding watershed. The day-long tour focused on illustrating existing problems, highlighting successful solutions, and fostering dialogue among the participants. Both the round table discussions and technical tour included invited participants from academia, local and federal government, and the private sector.

The first two watershed management conferences, hosted by the University of Virginia in Charlottesville, were held in 1997 and 1999. Numerous water resource professionals from academic and government agency backgrounds attended those meetings. This third conference, hosted by the National Taiwan University, brought the private and public sector together for detailed and meaningful discussions of an integrated approach to managing water quality and quantity in a sustainable manner.

Acknowledgments

Many people were responsible for the for the success of the Conference, including Conference Chairmen:

Jan-Tai Kuo, Hydrotech Research Institute, National Taiwan University
Shaw L. Yu, Department of Civil Engineering, University of Virginia

Conference Steering Committee:

Jan-Tai Kuo, Hydrotech Research Institute, National Taiwan University
Shaw L. Yu, Department of Civil Engineering, University of Virginia
Michael Clar, Ecosite, Inc.
Richard Field, U. S. Environmental Protection Agency
Edwin E. Herricks, University of Illinois at Urbana-Champaign
Michael A. Ports, HNTB Corporation
Nian She, Seattle Public Utilities

The Third International Conference on Watershed Management was organized by the Hydrotech Research Institute of the National Taiwan University, the Civil Engineering Department of the University of Virginia, and the Water Environment Research Center of the National Taipei University of Technology.

Major sponsors of the conference from Taiwan included:

National Taiwan University
National Science Council
Ministry of Education
Environmental Protection Administration
Water Resources Bureau of the Ministry of Economic Affairs
Water Conservancy Agency of the Ministry of Economic Affairs
Environmental Protection Foundation
International Commission on Irrigation and Drainage
Water Environment Research Center, National Taipei University of Technology
Council of Agriculture
Construction and Planning Administration, Ministry of Interior
US-Asia Environmental Partnership/American Institute in Taiwan
Chinese Institute of Civil and Hydraulic Engineering
Chinese Institute of Environmental Engineering

Major sponsors of the conference from the United States included:

Environmental and Water Resources Institute
Urban Water Resources Research Council
University of Virginia
Environmental Protection Agency

The knowledge shared through the formal presentations, facilitated discussions, and technical tour will lead to benefits realized far into the future. The engineering community is privileged to have had a dedicated group of people committed to the success of the conference. A special thanks is extended to all of the presenters for sharing their insight and experiences with their colleagues.

It should be noted that the pre-conference proceedings were edited by Jan-Tai Kuo and Shaw L. Yu and published in Chinese. The pre-conference proceedings are available directly from the Hydrotech Research Institute, National Taiwan University. These post-conference proceedings include the same technical papers translated into English as well as summaries of the facilitated round table discussions. All of the technical papers included in the post-conference proceedings are eligible for discussion in the Journal of Water Resources Planning and Management. Further, the papers are eligible for appropriate ASCE and EWRI awards.

PREFACE

The objective of the Third International Conference on Watershed Management is to promote the "integrated" approach to managing the water quantity and quality in water supply source watersheds so that the water resources in the watershed can be utilized in a sustainable manner. Problems of water shortage and water pollution exist throughout the world, including Taiwan. As we enter into the twenty-first century, the sustainability of water resources will play a critical role in continuing the future development for Taiwan.

The first two watershed management conferences were hosted by the University of Virginia, Charlottesville, Virginia, USA in 1997 and 1999. Many water professionals from academia and governmental agencies attended those meetings. The Third Conference is being hosted by the National Taiwan University in Taipei, Taiwan. More than twenty water management experts from USA, Japan and Taiwan are invited to present papers. The subjects of the presentations include those of critical importance to Taiwan, such as new technologies for erosion/sediment control; nonpoint pollution reduction, and hazard mitigation such as debris flow management, etc.

The Conference was organized by the Hydrotech Research Institute, National Taiwan University; Civil Engineering Department of the University of Virginia; and the Water Environment Research Center, National Taipei University of Technology. Major sponsors of the Conference include: from Taiwan, the National Science Council, the Ministry of Education, the Environmental Protection Administration, the Water Resources Bureau of the Ministry of Economic Affairs, and the Water Conservancy Agency of the Ministry of Economic Affairs; from USA, the Environmental and Water Resources Institute (EWRI), American Society of Civil Engineers (ASCE), and the Urban Water Resources Research Council (UWRRC), ASCE.

Co-sponsors of the Conference include: (all from Taiwan) Environmental Protection Foundation; Chinese Taipei Committee, International Commission on Irrigation and Drainage (ICID); Council of Agriculture; Construction and Planning Administration of the Ministry of Interior; Chinese Institute of Civil and Hydraulic Engineering; Chinese Institute of Environmental Engineering; Sinotech Foundation for Research and Development of Engineering Science and Technologies; and the China Times Cultural and Education Foundation. Many others made contributions to the Conference and all are gratefully thanked for their assistance.

Finally, we would like to express our special thanks to all the invited speakers and speakers of the Conference who contributed to the success of the Conference. We would also like to thank the Advisory Committee and other members of the Organizing Committee for their assistance in making all the preparation for the Conference.

Jan-Tai Kuo

Shaw L. Yu

CONTENTS

Research in Urban Stormwater BMPs

Daniel Sullivan[1] and Michael Borst[2]

Abstract

The use of best management practices (BMPs) in an urban watershed can provide adequate degrees of treatment at a relatively low cost. BMPs can range from being management operations (such as street sweeping or reducing the amount of pesticides used on urban lawns) to structural treatment options (the four most common alternatives being detention/retention ponds, swales, filter/buffer strips and constructed wetlands). This paper focuses on structural BMPs and overviews the state of the knowledge, the unknowns, and research programs being undertaken by the U.S. Environmental Protection Agency and other key organizations to address the unknowns.

Keywords: Best management practice; BMP; model; design; aquatic stressors; cost; imperviousness; hydrology.

Introduction

For the past three decades, municipalities in the United States have successfully addressed pollution in the watershed by collecting and treating their wastewater. Currently, all municipalities provide secondary level treatment, and in some cases tertiary treatment, and industries provide best available/best practicable treatment. This has had great benefits. More rivers are meeting water quality standards, and the public health is being protected from waterborne disease. The challenge now facing us is to address pollution associated with stormwater runoff, or wet weather flow, since this is now the last major threat to water quality. An associated challenge is in sustainable communities, that urban areas can continue to flourish while maintaining the basic necessities to support themselves.

It is less costly to prevent the generation of runoff than to treat it. Today, many municipalities are looking at low cost best management practices (BMPs)that prevent runoff. The lowest cost BMPs, termed non-structural BMPs, include practices such as limiting pesticide use in agricultural areas or retaining rainwater on residential lots (currently termed "low impact development"). There are a set of higher cost BMPs, which involve building a structure of some

1. Branch Chief, U. S. Environmental Protection Agency, National Risk Management Research Laboratory, Water Supply and Water Resources Division, Urban Watershed Management Branch, Edison, NJ 08837, USA
2. Chemical Engineer, U. S. Environmental Protection Agency, National Risk Management Research Laboratory, Water Supply and Water Resources Division, Urban Watershed Management Branch, Edison, NJ 08837, USA

kind to store stormwater until it can be bled into existing wastewater treatment facilities, that are also used. These can be more costly, especially in areas where land costs are high. The four most commonly used structural BMPs are detention/retention ponds, swales, filter/buffer strips and constructed wetlands; these are the focus of this paper and of the research currently underway at the Urban Watershed Management Branch of the US EPA Office of Research and Development.

In urban/suburban areas that are sewered, storage of WWFs is often used. Several cities use the option of storing flows in the large sewers throughout the metropolis. Modern approaches use computer technologies to track the storm, identify areas of high intensity, and adjust sluice gates/etc to store flows where the greatest capacity exists. These real-time control (RTC) methods have the advantage of using existing storage capacity rather than having to build separate facilities, and therefore are lower in overall cost.

For the past ten years, the Environmental Protection Agency (EPA) has been encouraging that water pollution controls be approached on a watershed basis. This is sometimes difficult due to municipal interests, but a watershed approach allows tradeoffs between point source treatment, pollution prevention, and optimal balances between the two. BMPs fit well within the watershed approach.

Current Concerns with BMP Performance

The overriding concern with structural BMPs is that you cannot link their installation to water quality improvements. In fact, receiving water quality often seems unchanged before and after the construction of the BMP. Two other major concerns are that we cannot tell what levels of pollutant removal can be associated with a particular BMP and that identical designs produce differing performance levels at different locations. Finally, there are few methodologies/models to tell water quality managers where to place the BMP in the watershed to get optimal water quality results.

Concerns about BMP performance leads us to a research program that addresses how the BMPs work, what they cost, how effective they are, and where to best place them in the watershed.

The five aquatic stressors of concern in the United States are nutrients, suspended solids and sediments (SSASs), pathogens, flow, and toxic chemicals. These same stressors have worldwide significance.

Current Concerns with BMP Design Guidance

There are several publications available that address BMP design. These publications are generally not well done and do not use basic hydrologic, hydraulic, and pollutant removal engineering approaches resulting in BMP facilities that are either not well designed or not well maintained. In addition, many practicing engineers for small municipalities do not incorporate water quality aspects of urban runoff, or small storm hydrology and its control into their stormwater control planning and design. As a result, many existing facilities do not perform properly and actually produce nuisance mudholes and standing water. BMP designs should address: (1) hydrologic criteria, such as continuous wet- and dry- weather infow, average storm characteristics, etc.; (2) hydraulic characteristics, such as flow velocities inside the pond or water retention times/ overflow rates in the pond, and (3) water treatment criteria, such as suspended solids (SS) removal, particle settling velocities, and nutrient absorption. Because of the increase of BMP usage and the current inadequate design guidance, there is a need for more complete guidance based on solid engineering principles.

Problem Statement

As a national assessment of urban runoff quality in the United States, EPA's Nationwide Urban Runoff Program (NURP) study[1] clearly showed that urban storm-water runoff was not clean or pure. Subsequent studies showed similar results.[11] The U.S. Geological Survey (USGS) further studied storm-water runoff,[111] again concluding that storm water carried a significant contaminant load. In short, urban storm-water runoff contains environmental contaminants at higher concentrations than found in the runoff from undeveloped areas.

Stream flow hydrology is significantly different in urban areas than in natural undeveloped areas. The major problem is imperviousness. Man-made impervious areas decrease rainwater infiltration and have a twofold effect on stream flow. First, because impervious areas quickly and efficiently transport rainwater away from an area, the stream flow hydrograph is sharpened and narrowed. Second, the decreased infiltration reduces available groundwater resulting in decreased base flow. Combined, these effects produce highly variable stream flow conditions, changes in stream power, and upsets to the established sediment equilibrium (forcing geomorphic alterations leading to stream incising and braiding). Unchecked urbanization increases runoff, decreases stream base flow, and destabilizes the streambed and bank. Booth and Reinelt,[1] studied forested watersheds of the Pacific Northwest and observed that when the 2-year post-development flow exceeded the 10-year pre-development flow in channels, the channels in the watersheds became unstable. This occured when the watershed imperviousness reached about 10%. Investigators from various humid U.S. regions support this observation, reporting both channel instability and an abrupt decline in indices of aquatic ecosystem integrity occurring at 10 to 20% watershed imperviousness.[11123] Other researchers,[4] directly measuring ecosystem indices, suggest the threshold values may be lower, perhaps as low as 5% imperviousness. Ongoing EPA-funded research suggests using a simple imperviousness threshold is an oversimplification. The connection between imperviousness and the distribution of rainfall within a watershed significantly affects the aggregated runoff.[5] Similarly, the local terrestrial characteristics (steepness, soil structure) and stream characteristics (slope, bed and bank geology, and vegetation) greatly influence the probable changes in fluvial geomorphology and resulting habitat alteration.[67]

Recognizing that increased runoff volume and increased contaminant concentration combine to generate an increased runoff load, it is not surprising the EPA's most recent release of the compiled States' biennial reports on water quality to Congress[8] showed the States collectively believe about 40% of assessed U.S. streams, lakes, and estuaries do not support the criteria for locally designated uses such as fishing and swimming. The commonly reported stressors include bacteria, nutrients, toxics (largely metals and pesticides), and silt. High stressor concentrations in storm-water runoff from agricultural and urban areas are a leading cause in the failures to meet designated use criteria. EPA analyzed the environmental effects of storm water[9] and urban hydrology,[10] concluding that the runoff from human-modified land areas can harm surface water resources, change flow patterns and hydrology, destroy habitat, and elevate pollutant concentrations.

Research Questions

The overall goal of EPA research is to develop techniques and approaches that allow community-based environmental planners to select cost-effective, watershed-wide solutions to restore and protect the receiving water quality from storm-water runoff contaminants on a watershed scale within a predictable time.

The basic research questions associated with BMP design and effectiveness are presented below.

Table: EPA-ORD Research Questions for Urban Stormwater BMPs

BMP Design/Costs	What is the best approach to selecting or eliminating from consideration the various BMP options?
	What are appropriate design requirements (e.g., removal of pollutants, quantity control, control of maximum discharge rates) of BMPs that will lead to achievement of management goals (e.g., protection of water quality, preservation of aquatic ecosystems)? What are the appropriate requirements of some stormwater control regulations (which determines BMP selection and sizing on the local level)?
	What design methods will result in BMPs that meet design requirements? How will design basis (hydraulic, hydrologic, particulate/dissolved fraction characteristics, water quality, stressors [or pollutant] reductions, etc.) influence BMP design and performance?
	How does cost influence design of BMPs? How does design influence construction and maintenance costs?
BMP Performance/ Effectiveness/ Placement in the Watershed	Does this BMP meet design performance requirements? What are the appropriate sampling methods (e.g., annual average vs. event-based measurement, concentration-based or influent loading-based removal for constituents, large vs. small storm quantity control, particulate/dissolved fraction characteristics, water quality, stressors [or pollutant] reductions,) for measuring BMP performance?

	Do BMPs continue to meet design performance requirements over time? Does maintenance impact BMP performance over time? What is the useful life of various BMPs?
	What are the ancillary benefits of various BMPs (e.g., aesthetic value, recreational space, multi-uses, wildlife habitat, irregation)?
	How do the performance of individual BMPs affect the watershed as a whole? Does BMP placement in the watershed matter?
	What are the costs of maintaining BMPs? How should BMP costs be evaluated?
BMP User-Assistance Tools	How can available models for design and analysis of BMPs (including computer simulations, design equations/models, and costing equations/models) be improved?
	Are new models needed for BMP design and analyses? What characteristics of a model are important for BMP design and analysis?
	Are current models adequate to apply for a system- or watershed-wide approach?
	How well do available models represent actual BMPs or watersheds with BMPs?

Source: November 2000. Internal Draft. Urban Stormwater Best Management Practices: a WWF Research Focus Area. by Richard Koustas, Environmental Engineer. USEPA, Edison, NJ.

Research Approach

Current watershed research in the branch is focusing on better definition of the BMP influent. Like any unit operation, the influent characteristic will strongly affect the performance. The collected data will simultaneously help to define the range that we must evaluate in the BMP model development. Despite national monitoring efforts, surprisingly few useful data are available to predict storm-water constituent concentrations. These monitoring efforts have produced a wealth of data identifying the constituents, but these data were largely collected for other purposes (e.g., defining if the water body meets the designated use) and generally do not fit our needs. EPA has begun research efforts to evaluate the expected storm-water concentrations using a nested hierarchical design to examine differences caused by land use and season. Follow-on work will establish if these results are regionally specific, and the extent to which they can be extrapolated to other regions and mixed land uses. Concurrent efforts are attempting to isolate the portion of the load that can be attributed to atmospheric deposition.

An early outcome of the ongoing problem-definition effort in describing the BMP influent is the clear evidence that existing permit compliance storm-water sampling techniques will be inadequate for research needs. Several issues, many mechanical, have developed during the EPA storm-water sampling. A simple example is the difficulty of measuring and sampling the low storm-water runoff flow rates, typical early and late in most storms and low-intensity rain events, in medium to large-diameter pipes.

The second stage of the research will occur at several physical scales. In some cases, the removal process is already reasonably well understood and requires only limited study (e.g., infiltration). In other cases, it is less clear that existing models can adequately describe the process in the needed techniques (e.g., temporal and spatial changes in infiltration rates with the accumulation of settled material in detention basins). Similarly, a physical process can apply to several different BMPs (e.g., chemical stressor partitioning between sediment and aqueous phases). Sometimes a series of experiments will be required. The coliform decay process is often described by a series of first-order relationships with time[11] based on light, temperature, dissolved oxygen, pH, and mechanical processes (sorption, filtration, and sedimentation). The EPA research will evaluate the first-order kinetic assumptions and estimate the kinetic constants generating a simplified combined decay model. Before beginning this work, we will measure the range of temperatures to be expected and the light levels, particularly the UV-B components from 280 to 320 nm as extinction coefficients with turbidity in local ponds. These data will help enable us to span the needed ranges in controlled-condition research.

Bench- or beaker-scale experiments should be sufficient to model many processes that occur in commonly used BMP. For example, this scale is adequate to augment the settling velocity data presented by Driscoll, et al.,[12] or to evaluate the equilibrium contaminant-sediment partitioning relationships. Other evaluations require larger-scale experiments. Relating the BMP residence time and dead zone formation,[13] a controlling feature in many processes (primarily in wetlands, detention, and retention ponds), to geometric shape and inlet position, to link hydraulic efficiency with clean sediment removal as suspended solids reductions using tracer techniques shows promise[14][15] but must be evaluated at physical scales greater than bench-top evaluations allow. This effort can build from the existing research.[16] Evaluating microbial processes in the rhizosphere similarly requires larger-scale experiments and longer experimental times. Generally, these require a test chamber that covers several square meters and will be completed in the on-site greenhouse. Finally, we require full-scale evaluations to properly consider previously controlled random events (e.g., wind) and evaluate the predictive ability of developed models. These evaluations will require measurements at constructed BMPs. Although we cannot identify a definitive list of the most commonly installed urban BMPs, patterns emerge when reviewing state regulations. The key selections appear to be retention and detention basins, constructed wetlands (including bioretention techniques such as rain gardens), infiltration techniques, and vegetated swales. WMRT storm-water research will begin with these BMPs, recognizing that we can often extrapolate the information to other BMPs. The increased hydraulic roughness measured in a swale will have obvious application to stream corridor buffer strips, for example. As additional information becomes available, the targeted BMPs may change.

Throughout the research, the pragmatic part of BMP construction must be a consideration, including retrofit concerns. Specifying grasses for swales, for example, must consider drought tolerance, mowing height and frequency, and the ability to slow overland flow. The water volume that must be controlled, the balance between extended holding to allow for settling and mosquito control, or BMP performance in low-volume rain events and flood events are other pragmatic concerns. New York specifies water quality management BMPs based on complete control of a ½-inch rain event,[17] the same volume Florida requires for drainage areas less than 100 acres.[18] Mosquito control, for example, suggests that the upper limit of extended detention must be less than the time required for mosquito development although longer detention would increase settling. Michigan recommends 48-hour holding and suggests less than 72-hours.[19] California recommends[20] 40-hour detention to capture fine clay particles while simultaneously cautioning against the associated short draw-down period linked with smaller storms having the same outlet configuration.

The broader question, restoration or protection of water quality, is much more complicated than the already substantial question of BMP performance. If we treated the total storm-water discharge to an impaired receiving water to pristine standards, it is unlikely that the receiving water would immediately recover. A time lag while residual contaminants leach from the sediment and habitat recovers, allowing benthic and higher order populations to return is part of the process. Community watershed planners need to have a realistic understanding of the time required for this recovery. Storm water will increasingly become part of the load considered in developing TMDLs. In-receiving water processes drive the recovery and attenuation process. To understand these issues, develop BMP effluent guidance, and define the needed predictive accuracy will require an ORD-wide effort.

References

1. U.S. EPA. (1983). Results of the Nationwide Urban Runoff Program, Volume 1 – Final Report.

2. Makepeace, D. K., D. W. Smith, and S. J. Stanley. (1995). Critical Reviews in Environmental Science and Technology. 25(2):93-139.

3. Marsalek, J. (1990). Evaluation of Pollutant Loads from Urban Nonpoint Sources, Water Science Technology. 22(10/11):23-30.

4. Driver, N. E., M. H. Mustard, R. B. Rhinesmith, and R. F. Middleburg. (1985). U.S. Geological Survey Urban Storm Water Data Base for 22 Metropolitan Areas Throughout the United States. Report 85-337. USGS: Lakewood, CO.

5. Mustard, M. H., N. E. Driver, J. Chyr, and B. G. Hansen. (1987). U.S. Geological Survey Urban Stormwater Data Base of Constituent Storm Loads: Characteristics of Rainfall, Runoff, and Antecedent Conditions;and Basin Characteristics. USGS Open File Report 987-4036. USGS: Denver, CO.

6. Driver, N. E. and G. D. Tasker. (1990). Techniques for Estimating Storm-Runoff Loads, Volumes, and Selected Constituent Concentrations in Urban Watersheds in the United States. USGS Water Supply Paper 2363. USGS: Denver, CO.

7. Booth, D. B., and L. E. Reinelt. (1993). "Consequences of urbanization on aquatic systems – measured effects, degradation thresholds, and corrective strategies." In Proceedings of Watershed '93. EPA 840-R-94-002. 545-550.

8. Morisawa, M., and E. LaFlure. (1979). "Hydraulic geometry, stream equilibrium and urbanization." In Adjustments of the Fluvial System. (Rhodes, D. D., and G. P. Williams, Eds.). Kendall-Hunt: Dubuque, IA. 333-350.

9. Booth, D. B. (1990). "Stream channel incision following drainage basin urbanization." Water Resources Bulletin. 26(3):407-417.

10. Booth, D. B. (1991). "Urbanization and the natural drainage system-Impacts, solutions and prognoses." Northwest Environmental Journal. 7(1):93-118.

11. Booth, D. B., and L. E. Reinelt. (1993). "Consequences of urbanization on aquatic systems – measured effects, degradation thresholds, and corrective strategies." In Proceedings of Watershed '93. EPA 840-R-94-002. 545-550.

12. Schueler, T. (1994). "The importance of imperviousness." Watershed Protection Techniques. 1(3):100-111.

13. May, C. W., E. B. Welch, R. R., Horner, J. R. Karr, and B. W. May. (1997). Quality Indices for Urbanization Effects in Puget Sound Lowland Streams. Technical Report 154. University of Washington, Water Resources Series.

14. Bledsoe, B. and C. C. Watson (in press) Effects of Urbanization on Channel Instability. Journal of the American Water Research Association.

15. Bledsoe, B. and C.C. Watson (2000). Logistic analysis of channel pattern thresholds:

meandering, braiding, and incising. Gormorphology 978

16. Bledose, B (2000) Stream Erosion Potential Associated with Differing Stormwater Management Strategies.

17. U.S. EPA. (2000). Water Quality Conditions in the United States: A Profile from the 1998 National Water Quality Inventory Report to Congress. Office of Water: Washington, D.C. http://www.epa.gov/305b/98report/98summary.html

18. U.S. EPA. (1992). Environmental Impacts of Storm Water Discharges: A National Profile. EPA 841-R-92-001. USEPA: Washington, DC.

19. U.S. EPA. (1997). Urbanization and Streams: Studies of Hydrologic Impacts: A National Profile. EPA 841-R-92-001. USEPA: Washington, DC.

20. Khatiwada, N. R., and C. Polprasert. (1999). Kinetics of Fecal Coliform Removal in Constructed Wetlands. IAWQ, Elsevier Science Ltd.

21. Driscoll, E. D., D. Di Toro, D. Gaboury, and P. Shelley. (1986). Methodology for Analysis of Detention Basins for the Control of Urban Runoff Quality. EPA 440/5-87-01. U.S. EPA: Washington, DC.

22. Werner, T. M. and R. H. Kadlec. (2000). "Stochastic simulation of partially-mixed, event-driven treatment wetlands." In Ecological Engineering. (14) 253-267.

23. Walker, D.J. (1998). "Modeling residence time in stormwater ponds." In Ecological Engineering. (10) 247-262.

24. Rash, J. K., and S. K. Liehr. (1999). "Flow pattern analysis of constructed wetlands treating landfill leachate." Water Science Technology. (40) 3, 309-315.

25. Werner, T. M., and R. H. Kedlec. (2000). "Stochastic simulations of partially-mixed, event-driven treatment wetlands." In Ecological Engineering. (14) 253-267.

26. NYSDEC. General Permit for Construction Activities– Stormwater Management Guideline for New Development (TOGS 5.1.10), Appendix D.

27. Florida DEP. (1995). Regulation of Stormwater Discharge. 62-65.

28. Michigan Department of Environmental Quality. (1999). Stormwater Management Guidebook.

29. California. (1993). Municipal Handbook.

Watershed Management Strategy in Taiwan

Nien-Ho Ma[1], Miehshan Huang[2]
and Kwai-Wei Fong[3]

Introduction

With the rapid urbanization and intensive industrialization in the past decades, the environmental quality in Taiwan has been greatly degraded in many aspects, such as water quality. Given the fact that river quality is a key element for national health as well as a pivotal indicator of living quality, it is an extremely important subject to strengthen the watershed management strategy as to resolve pollution problems and maintain sustainable aquatic environment.

Pollution Conditions

According to the monitoring data, it is indicated that the river quality reached the worst condition in 1996, during which period the percentage of severely polluted river sections was once as high as 14.9% as opposed to 12% for the time being. Although the severely polluted river section has been lessened in the past years, the improvement appears to be slow, especially for those urban rivers with dense population. The downstream areas of nearly 10 rivers in the southwest regions in Taiwan have been classified as "severely-polluted", which has imposed great threats on the living quality and drinking water safety for the local residents.

As for the water quality of reservoirs, monitoring data of 20 major reservoirs indicated that 15 reservoirs were eutrophicated in 1997, as opposed to 7 in 2000. The quality of many reservoirs has been improved in the past years, whereas some still remain polluted.

In brief, it is obvious that the water quality of aquatic systems in Taiwan has been improved as long-term pollution control projects being implemented in the past years. However, in order to meet the increasing demands on water quality from the general public, the Environmental Protection Administration in Taiwan (TEPA) is determined to carry out various projects to maintain the natural balance of the ecosystem and the sustainable development of the environment.

---

1. Director, Water Quality Protection Bureau, TEPA, Taiwan
2. Chief Engineer, Water Quality Protection Bureau, TEPA, Taiwan
3. Specialist, Water Quality Protection Bureau, TEPA, Taiwan

Implementation of River Clean-up Projects

1. Projects Based on Pollution Categories

1.1 Rivers of Industrial Pollution Type

(1) Removal of Illegal Metal-Refinery Factories along the Riverbank of Er-Zeng Brook

The Er-Zeng Brook was severely polluted by illegal metal-refinery factories through the acid-wash process of mixed metal scrap in 1987. This pollution was notorious for its well-known "green oyster" incident, which is believed to be the result of copper pollution. The pollution situation has been greatly improved as a result of strict inspection programs implemented by the environmental authorities. For the time being, the copper concentration in Er-Zeng Brook has complied with the water body standard.

In the June of 2001, a joint team of all environmental authorities swept 68 illegal factories out of the Er-Zeng Brook area, 11 of which were introduced to nearby industrial parks whereas the remaining 57 being torn down completely. The air and water quality in the area has been substantially improved since then.

(2) Industrial Wastewater Control with Reasonable Effluent Standards

With the domestic economic condition going downward lately, some industries would complain and blame on the strict environmental standards. However, among a number of factors that could affect the economic development, environmental standards alone should not be held responsible for such conditions. Nevertheless, the balance point between economic growth and environmental protection can still be re-considered based on the changing conditions around the world and inside the country as well. The Taiwan's TEPA will conduct relevant studies in determining the most appropriate and up-to-date effluent standards.

1.2 Rivers of Livestock Pollution Type

(1) Implementation of Prohibition and Compensation Project for Pig Farms in 5 Major Watersheds

There are about 7 million pig capitals in Taiwan, which impose some major water pollution on the river quality. In order to deal with wastewater from pig industries contaminated in the river stream, it is necessary to increase the disinfection dosage in the drinking water treatment processes. Therefore, the risk of carcinogenicity among the general public has been increased accordingly. To protect the drinking water safety in the water source zones, the TEPA started to implement the prohibition and compensation project for pig farms in 5 major watersheds, including Kao-Ping Brook, Tam-Shuin River, Tao-Chian Brook, Da-Giar Brook, and Tseng-Wen Brook. Approximately 640 thousand pig capitals are expected to be removed from these areas.

As of October 2001, nearly 96.7% of pig farms have been removed with compensation from these watersheds. It is believed that the drinking water quality for

12 million residents in the areas can be effectively improved, upon the completion of the entire project by December 2001.

(2) Implementation of the Kao-Ping River Clean-up Tasks to enhance the drinking water quality in the Kao-Hsiung area

The drinking water quality in the Kao-Hsiung area has been criticized for a long period of time. Beside the implementation of the prohibition and compensation project for 510 thousand pig capitals in the upstream area of the Kao-Ping River, other projects such as the construction of urban sewer systems in Chi-Mei, Da-Sue, Kao-Sue, and Ping-Tung City are also undertaken to resolve the domestic wastewater pollution problem. Once these entire construction projects are completed, it is anticipated that the NH3-N concentration can be greatly reduced from 4.5 mg/L to 0.66mg/L, or an 85.4%reduction. Besides, the drinking water treatment plants in Tsen-Chin Lake, Kow-Tan, and On-Kun-Yuan will also be upgraded with new technology, in order to thoroughly improve the drinking water quality.

(3) Promotion of Low/Zero Pollution in Pig Farms beside the Water Source Zones

Beside 5 major water source zones where the prohibition and compensation policy have been implemented, various low/zero pollution technology and management strategy have also been promoted in pig farms beside the water source zones. The pig farm industries have long been criticized for their high levels of pollution, especially when most pig farms are not equipped with effective and efficient treatment facilities to remove major pollutants such as nitrogen (N) and Phosphorus (P). The key solution to resolve the problem is to adjust the basic raising pattern in pig farms. For instance, the idea of "pig-toilet" has been successfully introduced to some pig farms in Taiwan. Besides, the sTEPAration of pig manure and urine would also help reduce the water pollution. Some of this solid manure can be well incinerated for recovering energy, or can be composted to generate useful organic fertilizers. Most of these processes would reduce the chance of water pollution, and in consequence would impose less pollution in the environment.

1.3 Rivers of Domestic Wastewater Type

(1) The Tam-Shuin River Clean-up Project

To protect the drinking water safety and provide recreational spaces for 6.6 million residents in the Taipei Metropolitan Area, the TEPA started a 10-year massive clean-up project in the Tam-Shuin River Watershed in 1987. A follow-up project was initiated in 1998. The goals of these projects include the no-trash condition above the water surface, the maintenance of water quality in the source zones, and to provide additional recreational spaces.

The clean-up project has produced some major achievements. The joint Pa-Li sewer system was successfully test-run on June 28,1997, with an operation rate of 900 thousand tons of wastewater each day. It is expected to reduce 30% of water pollution in the watershed when the operation rate reaches its full capacity of 1.32 million tons of wastewater per day.

For the time being, monitoring data around the watershed indicate that the average dissolved oxygen concentration in the entire river bodies has reached its goal of not-lacking-oxygen. Fish species are also increasing in the past years. Besides, 46 recreational areas or green-belt parks have been established along the river bank, which in total provides nearly 1000 hectares green space for the public.

(2) The Pu-Tzi Brook Clean-up Project

The Pu-Tzi Brook in the Chia-I area is one typical river of mixed pollution source from industrial, livestock, and domestic wastewater. The TEPA started its major clean-up project in 2001 with relevant strategies, including the construction of sewer treatment facilities, the enforcement of inspection on riverside industries, the garbage clean-up along the riverside as well as on the water surface, the establishment of highland ecological area along the riverside, and the promotion of river environmental education. The expected benefits, upon the completion of these projects, include the increase of water quality compliance ratio from 28% to 60%, the protection of drinking water safety in the upstream area, the development of bio-diversified aquatic environment, and the safety of aqua-cultural industries near the joint point into the ocean.

2. Enhancing the River Environmental Quality and Creating Recreational Space

The TEPA started the River Watershed Management Project with the clean-up projects for Er-Zeng Brook and Gion-Ging Brook in 2000 and those for Bei-Guan Brook and Tseng-Wen Brook in 2002. The main issues of the projects are to clean up the polluted spots along the riverside, and to improve the river ecosystem. Major accomplishments include the provision of highland green areas along the river sides, and 6 fish-pond eco-garden in the wetland. A total of 30 hectares of recreational areas are provided for the general public. In addition, similar projects are to be carried out for the Chien-Tseng River the Kaohsiung City.

Other major strategies in these projects include the strict enforcement of site investigation along the riverside, the prevention and control of pollution in the watershed, the enhancement of water quality in the upstream reservoirs, the management of highland wetlands, and the improvement of domestic sewer in the nearby urban areas. Overall speaking, a total of 30% of water pollution is expected to be removed by these projects.

3. Management of Non-point Source Pollution

The operation manual of best management practices for pollution prevention and control in the industrial parks, recreational areas, agricultural zones, and construction sites has been edited and published to be used as the basis of controlling non-point source pollution.

4. Imposition of Water Pollution Control Fee and other Economic Incentives

Beside the stringent enforcement of law, the establishment of economic incentive system is also considered as one effective tool. Among various economic incentives, the imposition of water pollution control fee appears to be the second most effective measure, only next to monetary fines. The basic principle of imposing such fees is to substantiate the idea of "polluters pay", and to reduce the emission or discharge of pollutants through economic mechanism. The imposition of such fees can benefit in at least two aspects. On one hand, this will motivate the industries to enhance their wastewater treatment facilities, and consequently reduce their pollution. On the other hand, this imposition will stimulate the development of environmental protection industries, such as consulting firms, engineering services, and R&D sectors, all of which are believed to be able to help boost the economy. In this sense, the imposition of water pollution fee carries out the active meaning of pollution prevention. Some countries in Europe, such as Netherlands, Germany and France, have adopted similar systems, and their pollution indexes have shown obvious decreases with the growth of economic indexes.

In Taiwan, the legal basis of imposing water pollution control fee has been clearly stated in the water pollution control Act. However, the imposition of such fees has not yet been implemented as of now because of some problems, such as the identification of the imposition targets and authority levels. The amendment to the Act has been proposed to the Legislative Yuan, and the TEPA will continue working on the establishment of all legitimate bases required to implement such imposition.

5. The Need to Increase of the budget and Manpower to Implement Water Pollution Clean-up Projects

With the growths of modern development and dense population, the pollution problems have become more complicated, which accordingly enhance the difficulties of dealing pollution control. In addition, mobile source of pollution, such as illegal dumping of domestic and industrial wastes or even hazardous materials, has become more often in the river pollution. This situation has certainly paralyzed, at least in part, the limited manpower in the authority. Besides, with the limitation under current financial conditions, the budgets for clean-up projects are obviously inadequate. For example of the 23 river clean-up projects proposed by the TEPA, a total amount of 40 billion NT dollars is needed for a ten-year term, without mentioning additional costs for constructing sewer systems which are classified as the duty of the Interior Ministry. Yet the real case is that the available budget for the clean-up projects in 2001 is only about 0.23 billion NT dollars, which is far lower than proposed.

Besides, the budgets for local environmental authorities are also running short. Some local governments are even experiencing financial crises after the adjustment of budget distribution patterns from the central government. It is indeed a major challenge for Taiwan's TEPA to achieve the goals of pollution control with very limited budget and manpower.

Concluding Remarks and Future Perspectives

It is by all means the duty of TEPA to provide the general public with safe drinking water and quality environmental amenity. Since the establishment of Taiwan's TEPA in 1987, the population in Taiwan has increased by 13%, whereas the numbers of factories and pig capitals have been increased by 17% and 5%, respectively. The environmental burdens are greatly increased in accordance. Although some achievements in water quality protection have been perceived, it is still the wish of Taiwan's TEPA to provide a better living environment with safe drinking water, bio-diversified ecosystem, adequate recreational areas, and sustainable development.

Watershed Management and Protection – Issues of Scale and Process in Ecosystems

Edwin E. Herricks[1]

Abstract

Although watershed management science and practice is not new, there has been an emphasis in the latest revival of watershed management programs on ecosystem protection and restoration. There has been a shift from protecting society from the environment to protecting the environment from society. With this shift, criteria to measure programmatic success has shifted, placing ecological outcomes of management programs and practices high on the list of watershed management objectives. The integration of ecosystem-based objectives in watershed management programs demands a substantial shift from past practices. First, watershed management must consider temporal and spatial scales in an ecological context. In this ecological context, temporal scale is associated with the life history of organisms, which may be affected by evolutionary history that equips organisms to function in a given environmental setting. Spatial scale is associated with a set of nested scalar relationships that span the range from individual organism size and movement capacity to global scales that reflect the influence of transient or colonizing organisms. Second, watershed management, directed to ecosystem protection, must consider systems-level processes in ecosystems. A systems view recognizes an organizational hierarchy of processes that start at the organism level, finding emergent characteristics at each level of organization. The implementation of ecosystem scale and process elements suggests a new paradigm for watershed management. This paradigm integrates past management science and practice with ecosystem protection based on ecosystem-relevance filters, which assist in the development of projects that produce needed ecosystem benefits.

Introduction

The objectives of watershed management have changed as the emphasis has shifted from the management of damaging floods and the development of reliable water supplies to the protection of watersheds and their associated ecosystems. The early emphasis depended on engineering solutions and the application of structural controls. The result was the construction of large and small engineering projects that replaced natural conditions with ordered, highly controlled facilities and the corresponding degradation, if not destruction, of local and regional ecosystems. There has been a shift away from watershed management projects that consider only engineering objectives. This

[1] Professor of Environmental Biology, Department of Civil and Environmental Engineering, University of Illinois at Urbana-Champaign, 3230b NCEL, MC-250, 205 N. Mathews, Urbana, Illinois

shift recognizes the fundamental importance of sustaining ecosystem function. Although some ecosystem functions may require maintenance of a completely natural setting, both engineering practice, and our understanding of ecosystem structure and function, can be used to optimize projects to support sustainable ecological systems while meeting management objectives.

Watershed management, and the associated assessments of the performance of management programs, face a new operational paradigm. This new paradigm is exemplified in the United States Environmental Protection Agency's Watershed Protection Approach (WPA). The WPA provides a watershed-orientation to meeting water quality goals, which takes into account all threats to human health and ecological integrity within specific watersheds (EPA, 1993, EPA 1995). To some extent, this approach is a departure from a focus on regulating specific pollutants and pollutant sources in that the WPA integrates traditional regulatory and non-regulatory programs to support natural resource management. Possibly the most important foundation for the WPA is the coupling of partnerships, a geographic focus, and sound management techniques based on strong science and data. Partnerships are intended to ensure that watershed management objectives integrate economic and other social and cultural goals. Further, partnerships provide public input to all stages of watershed management program planning and implementation. The geographic focus provides a means to logically bound management by establishing logical connections due to surface drainage or groundwater influence. Finally, the WPA assembles all watershed stakeholders in a setting where sound scientific data, tools, and techniques can be used in an iterative decision making process. This includes:

- assessment and characterization of the natural resources and the communities that depend upon them;
- setting goals and identification of environmental objectives based on the condition or vulnerability of resources and the needs of the aquatic ecosystem and the people within the community;
- identification of priority problems;
- development of specific management options and action plans;
- implementation; and
- evaluation of effectiveness and revision of plans, as needed.

The WPA is not without its difficulties. Possibly the most troublesome change for engineers and environmental scientists is the changed role of technical experts in the process. In the WPA, technical experts interact at the earliest stages of project planning with stakeholders for all interests in the watershed. Technical experts no longer have absolute control over project designs. Technical expertise has lost its privileged position of authority and influence on final outcome and is often the same as any other stakeholder in the projects. Thus scientists and engineers can no longer expect to be the only authority, and watershed management requires an expanded iterative process leading to decision making. Public wants and needs often challenge traditional engineering solutions. This new management setting has produced a challenge for technical professionals while at the same time making greater demands for the rigorous scientific support for any project. In this new setting, the importance of ecosystem sustainability has grown so that protecting the environment is given equal weight, if not more emphasis, than protection of public values from environmental extremes.

Ecosystem Sustainability in Watershed Protection

To address the need to incorporate notions of ecosystem sustainability in watershed management, a starting point is the recognition that the environment, and associated ecosystems, are governed by fundamental physical, chemical and biological/ecological processes. These processes operate over a range of spatial and temporal scales and may be specific to a location in the watershed. For example, if we consider flow and water quality issues in a low-order stream reach, the changes in physical and chemical conditions, as measured from baseline/baseflow condition, will be rapid and of short duration. Although the duration of change may be short in low order streams, the magnitude of change, as compared to baseflow is large during flood events. Residual effects in low order streams include changes in channel form and stability, while the accumulation of contaminants is often limited because of the high rates of transport from low order streams. The ecosystems in low order streams will reflect the physical and chemical conditions of spatial location (Vannote, et al., 1980) with changes in species presence and abundance changing rapidly. To protect the ecosystems in low order streams, management will often focus on both riparian and channel issues. Riparian objectives will emphasize maintenance of allochthnous sources of carbon, and maintenance of suitable buffer zones to protect the channel ecosystems. Channel objectives will emphasize control of contaminant concentration, rather than loading, and the management of hydrograph change to promote high quality, and stable, physical habitats Herricks (1996) . Flow and water quality issues in higher order streams reflect the fact that the rate of change in physical and chemical conditions is slower than observed in low order streams, while the duration of change is longer, reflecting the concentration time needed for flows from tributaries. In high order streams water quality concerns often focus on loading rather than concentration because contaminants may be stored in sediments and produce localized high concentrations that may be toxic. The ecosystems in larger streams and rivers will also reflect the physical and chemical conditions of spatial location. Riparian objectives will include maintenance of floodplain integrity and connections between the channel and the floodplain. Channel objectives will emphasize control of loading or management of transport, and the maintenance of diverse habitat to meet multiple needs of species life histories. The management focus in larger watersheds must also deal with effects of multiple sources, and the effect that a more diverse land use has on both the number and mass loading of contaminants.

Integration of Toxicity Testing

It is possible to extend this analysis of the management focus for low and high order streams to predict ecological consequence from the measurement of toxicity. Although it is often difficult to extrapolate ecological consequence from the toxic response of a single species, it is possible to use a knowledge of this toxic response to help focus management on sources of contaminants and the practices needed to maintain stable habitat in stream channels. The starting point for this analysis is the recognition that toxicity is produced by the concentration of a contaminant, the duration of exposure, and the frequency with which an exposure is repeated.

The disturbance of any watershed, natural or anthropogenic, is expected to produce alterations in runoff volume, and chemical characteristics of the runoff (Borman et al., 1968; Vitousek et al., 1979; Webster and Patten, 1979; McDowell, 1988; Close and Davies-Colley, 1990; Edwards and Helvey, 1991). Unfortunately, the concentration of humans also leads to the concentration of contaminants that is associated with the proliferation of landscape/watershed disturbances both near the population center and at locations where resources are obtained to meet population needs (Goudie, 1990). The result is that water quality alterations are magnified by the combined effects of landscape disturbance, changes in rainfall/runoff relationships, and the presence of contaminants. Watershed use changes will produce different concentration, duration of exposure, and frequency of exposure conditions, which will differ based on stream size and watershed area. The effects on receiving system ecosystems will be short-term if the natural assimilative capacity of the stream/watershed system is not exceeded. As noted above, watershed location (stream order) is critical in determining the relative importance of concentration or loading. When low order streams are compared with high order streams, the effects of changes in runoff volume and contaminant concentration are less likely to be moderated by ambient flows leading to acute toxicity and profound, short-term change in receiving system biota. In high order streams contaminant concentrations may not be acutely toxic, but the residual effects produced by deposition, storage, or slow transport will lead to chronic toxicity and the effects on receiving system biota will be subtle. As the ratio of disturbed to undisturbed land increases, the risk of severe impact also increases. Recent analyses indicate that any increase in impervious area above 15 to 20% will severely limit ecological integrity (Shaver, et al., 1995).

This complex set of relationships among spatial scale, duration of effect, location, and ecosystem condition does require that watershed management use a new paradigm to "protect" ecosystems. We do know that we can use batteries of tests that operate over a range of scales (Herricks and Milne, 1998) to measure the effects of chemical contaminants and mixtures of contaminants. We can also use a range of techniques to identify the consequence of land use on flow and habitat (Blesdoe, 2001). What is necessary is the integration of this information to meet the needs of an ecosystem management paradigm. The foundation for this paradigm is an understanding of some critical systems-level issues applicable to ecosystem protection.

Critical Systems-level Issues

The appreciation of a systems view is a critical component of modern watershed management. A system is simply two or more components that interact (Hall and Fagen, 1956) where the delineation of the system may be arbitrary. Thus in a watershed system, boundaries are established by watershed characteristics, and include physical, chemical, and biological conditions and process as components. A systems view for ecosystems in this paradigm is founded on a hierarchical organization of system components where each level of the hierarchy is characterized by increasing complexity: Organism → Population → Community → Ecosystem → Biome. Next, it is recognized that system size and complexity will often control opportunities for detailed analysis of process, that a disturbance to process at one level of the hierarchy may be unrelated to or undetectable at another level of the hierarchy, and that temporal and spatial scales may be tractable at lower levels of the hierarchy (species and population levels), but intractable at and above that level. Finally, this view recognizes ecosystem organization and function is based on system elements or components (species) and their structure or organization into trophic levels in food chains or food webs, and system function and the dynamics of material and energy transport, which depend on the linkage and interaction strength among species in food chains and food webs. In fact, an accurate systems view requires the integration of system elements and processes with environmental conditions so that the linkages between landscape condition, physical and chemical process, habitat features, and biotic communities are well delineated and understood.

Other critical issues in this new watershed management paradigm arise from considerations of system organization and dynamics. It is critical to define the "normal" states of systems, and the expected rate at which watershed systems change through time. Further, the analysis, or simulation/ modeling of a system should extend for a length of time equal to the life history of the longest lived organism in the system to understand system dynamics. What this means is that true ecological integration may require a time span equal to one complete turnover of biotic elements (population numbers, biomass), which may extend 10's, 100's, and even 1000's of years.

The next set of critical issues address how we organize our thinking about the system. Although the identification of system elements and function is a first step, this new paradigm requires a thorough understanding of system operations, specifically cause and effect relationships between environmental change and ecosystem stability. One of the easiest ways to develop this operational sense is to apply a general model formulation approach that identifies input variables, establishes state variables for model algorithms, defines transfer functions, and produces an output that can be used to predict future conditions. Remember that this model formulation is designed to provide an ecosystems view so the input variables must be ecologically relevant, have an appropriate spatial and temporal scale, and operate as dependent variables in the cause and effect relationships captured by the state variables and transfer functions (Herricks 1996). For state variables, it is critical to identify ecosystem properties or processes that are the core of the modeling effort. Critical ecosystem properties or processes include elemental dynamics, energy dynamics, food web (trophic dynamics), biodiversity, critical species (presence/abundance), genetic diversity, dispersal and migration mechanisms, influences of natural disturbance, and maintenance of the capacity for ecosystem development (NRC, 1991). It is then necessary to consider transfer functions, where input and state variables are related and the mechanics of the model are defined. Finally, the output is designed to provide predictions of system element composition, system structure and organizational relationships, and system function and dynamics. In summary, one of the values of this model formulation approach to the development of a watershed systems view is that the model development process requires the development of an understanding of system organization and dynamics at scales appropriate to the model formulation. The approach also supports valuable "what if" analysis that includes identification of new input variables, redefinition of state variables and transfer functions, and development of output that will accommodate ecological interpretation. Finally, the approach provides a means to define a protocol that both guides new data acquisition and uses that data to continuously improve model formulation. The result is the development of an adaptive approach that can accommodate natural variability and the expected natural change that often confounds ecosystem-based management approaches.

Implementation of an Ecosystems View

The implementation of an ecosystems view in a watershed management program is presently underway. The U. S. Army Corps of Engineers, Rock Island District Office in Illinois and the Illinois Department of Natural Resources is cooperating on the Illinois River Ecosystem Restoration Program. The initial stages of this program were initiated in May, 2001 with an initial planning effort to be completed in 2004 and continuing project implementation through 2020. The scale of the program is the entire Illinois River watershed, Figure 1, with particular attention paid to main channel restoration and also the management of sediment and water quality sources throughout the watershed.

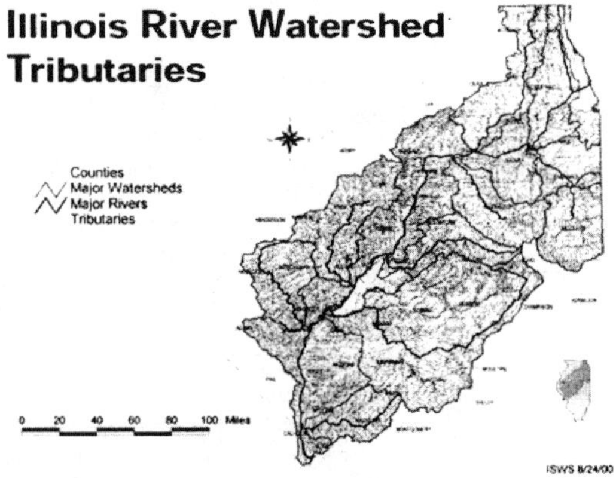

Figure 1. Illinois River Watershed with sub-basins identified.

The initial program activities have centered around a Restoration Needs Assessment (RNA). The RNA is the first step in the Corps of Engineers planning process. The RNA has several major tasks. An overall watershed assessment includes an information assessment, a geomorphic evaluation, a land use, and habitat evaluation. Assessments specific to the main stem of the Illinois River include an evaluation of sediment problems and hydrology, and a hydraulic evaluation focused on water level maintenance. A second task is a tributary assessment where tributary watersheds are defined and meetings with local stakeholders are planned to identify problem areas leading to a detailed description/characterization of the watersheds. A third task is directed to evaluation of sub-watersheds that is intended to prioritize watersheds for action, and support the development of specific projects.

In addition to the RNA, the Illinois Department of Natural Resources (DNR) is involved in a process to identify project needs and implement project activities to improve the general quality of the Illinois River and its tributaries. Projects include side channel and backwater area restoration to replace aquatic habitat diversity lost from 100 years of uncontrolled siltation, which has resulted in approximately 75 to 90% loss of the water depth. The projects are designed to remove accumulated sediment in selected areas and place rock and rubble substrate to create a diversity of depths and substrates, which can serve as habitat for different life stages of fish and other aquatic life. In addition to main channel projects, watershed projects that develop and implement various conservation practices which are either not currently included in existing state and federal programs or were not implemented due to resource availability in those programs are planned. The projects will be implemented with volunteer landowners (public and private) in targeted areas.

In reviews of the RNA, and the projects proposed by the DNR a questions were regularly raised about ecological relevance of both the RNA approach and the proposed DNR projects. The basis for the questioning was the lack of specific, and ecosystem-relevant, criteria for planning and project implementation. To meet the needs for ecological relevance review, an approach was developed that is designed to provide a systems-level test that has clear ecosystem relevance in the RNA and the DNR project performance assessment. To initiate the systems approach a defined ecosystem objective was established. Considering a wide range of national actions supporting management and regulation of human activities to protect the environment, and the past emphasis on restoration and maintenance of the physical, chemical, and biological integrity, the objective for the program was defined as restoration and maintenance of ecosystem integrity. Ecosystem integrity, sometimes related to as ecosystem health (Schaeffer. et al., 1988), is both a concept, and a quantifiable measure. As a concept, ecosystem integrity recognizes the need for integration of physical, chemical, and biological components of the environment, with a specific intent of restoring and sustaining ecosystems that are natural in character. As a quantifiable measure, ecosystem integrity is assessed in terms of system organization and dynamics. As might be expected from earlier discussions, characterizing, and quantitatively measuring, ecosystem integrity can be problematic if the dominant management approach is directed to individual species rather than the system/ecosystem.

The basic foundation for setting ecosystem integrity as a goal is the recognition that ecosystem integrity is the product of the influences of multiple factors in ecosystems. These multiple factors have been identified by the National Research Council as critical attributes of ecosystem integrity and sustainability (NRC, 1991). As stated in the NRC report "Although critical attributes are only components of ecosystems and landscapes, their status can be used to evaluate linkages that make an ecosystem function as a whole." The critical factors, and potential measurement variables appear as ecosystem integrity check points in the systems integration (Figure 2). The importance of critical attributes in this scheme is that they serve as check points, or the quality assurance element of the approach. Further, the selection of reasonable measurements for these attributes for the system under evaluation, can provide both a site-specific basis for evaluation and a systems-based evaluation.

In the application to the Illinois River Ecosystem Restoration Program, the critical attributes serve two functions. The first is a checklist for integrity evaluation. The second is a filter that assists in categorization and organization of information in the integrity analysis. As a checklist, the critical attributes function to remind the analyst that at the systems level, there are multiple conditions that must be met to assure ecosystem integrity. The recognition of multiple conditions encourages the analysis of connections between and among variables, and provides the operational approach to model formulation and testing. As a filter, the critical attributes support the simplification of the complexity of ecosystem organization and dynamics into categories that are most relevant to the project that must be evaluated. The filters, as conceived in this approach, are similar to the filters described by Poff (1997), which are scaled features that influence the probability that species with specified functional attributes are able to persist in a community of organisms. In the sense of the model formulation suggested earlier, the filters represent the combination of the input variable and the state variable and transfer function that is developed to follow cause and effect to a defined output. In Figure 2, the ecosystem integrity filters, have been grouped for convenience to address the critical attributes that are associated with logical ecological analysis endpoints for the Illinois River Ecosystem Restoration Program. The three endpoints encompass food and nutrient dynamic issues, species maintenance issues, and natural change issues. This initial grouping is intended to evolve as model formulation and future monitoring reveal both general and site specific issues that must be addressed. It should also be noted that a singular focus on single endpoints is not intended, as indicated in the figure, there is interaction between and among endpoints and identified endpoint measures.

Figure 2 also identifies a project parsing process to identify ecosystem relevance. This project parsing is intended to provide a means to more thoroughly evaluate project value to ecosystem integrity by identifying specific aspects of project performance applicable to critical ecosystem attributes. As in most watershed management programs, the Illinois River Ecosystem Restoration Program has had a primary focus on problem solution. A difficulty arises when the problem-oriented focus fails to provide a basis for establishing clear and valid connection to integrity restoration or maintenance. The argument is commonly made that "if we build it, they will come," implying that if the physical and chemical conditions are improved, organisms will recolonize a site and restore the ecosystem. Unfortunately, a project focus may not produce needed conditions, or address critical attribute needs. The parsing process is thus intended to provide a careful assessment of project value to ecosystem integrity. Rather than a filter, this parsing process may be viewed as a prism that takes the coherent and focused approach of the project and identifies a spectrum of possible relationships to ecosystem integrity. In the Illinois River Ecosystem Restoration Program, again, the categorization was based on convenience and considers project types, such as main channel projects, projects in the immediate riparian zone for large channels or along all lower order stream channels, and projects at locations beyond the channel or riparian zone that have a landscape focus rather than a channel/drainage network/riparian zone focus. As with the ecosystem integrity filters, it should be noted that a singular focus on parts of the spectrum is not intended. As indicated in the figure, there is an expected interaction between and among spectrum of relationships identified for any project. Further, the figure also indicates that there is an expected interaction between, and among, filter levels. This design encourages an "open" model formulation approach that has both general and site specific application.

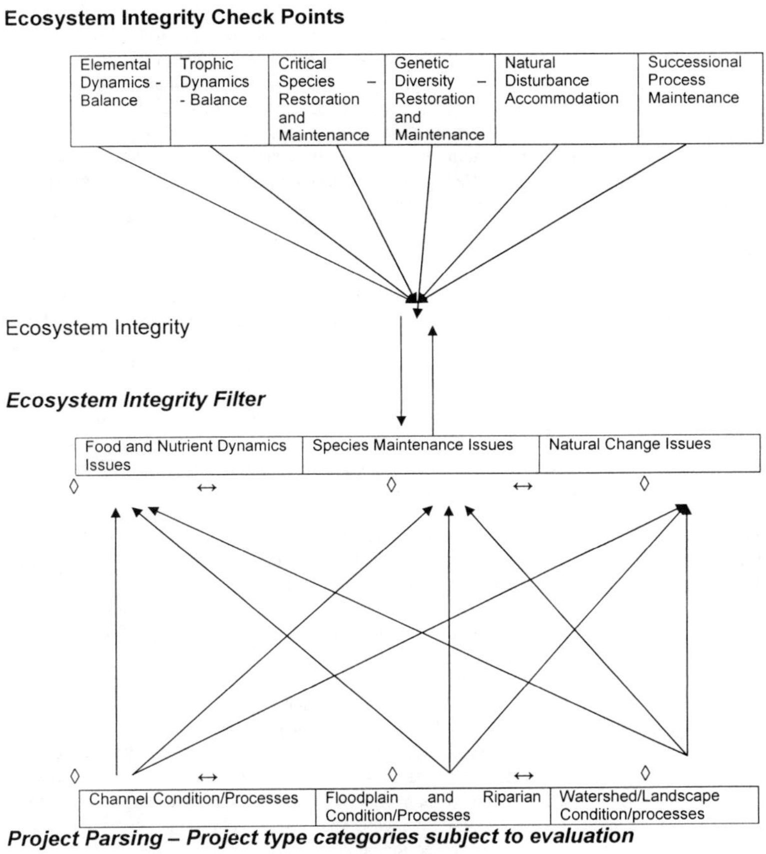

Figure 2. Proposed organization to relate project activities to ecosystem integrity in the Illinois River Ecosystem Restoration Program.

In summary, the proposed approach is intended to provide a means of organizing review processes, categorization, ecological valuation, prioritization, etc. in the Illinois River Ecosystem Restoration Program. This approach is not intended as a final, or all encompassing procedure, rather it is intended as a component of the overall program to help organize the difficult process of addressing ecosystem issues where program and project issues can be addressed within a framework of ecological integrity assessment, which is based on well developed thinking from ecological theory. The approach is not designed, or intended, to establish a hierarchy of relationships that lead to prioritization of integrity attributes that must be met, rather it is intended to "level" the playing field by simply identifying important associations and values of the program and projects when an ecosystem integrity goal is clearly identified. This approach is presently evolving as the Corps of Engineers and the Illinois DNR continues the planning and implementation of Illinois River ecosystem restoration projects.

Conclusions

The integration of a systems view in watershed management programs, in particular a systems view that includes a realistic ecosystem perspective, is possible, but unfortunately all too often elusive. The primary problem is that an adequate appreciation of scale is not a part of the project development process. In that development process, it is easy to locate a project and define individual project performance. It is much more difficult to determine if a group of individual projects are meeting a system need. This general problem is exacerbated when programmatic objectives are specific to ecosystem restoration and protection, and there is only a limited understanding of ecosystem needs, and an even more limited understanding about how to assure a desired ecosystem outcome can be produced.

The discussions in this paper have been designed to suggest that it is possible to capture ecosystem objectives in modern watershed management. The first issue that must be addressed is bounding the scale of a management effort, with the recognition of the need to connect site-specific measures with system-level needs. The next issue is the development of an understanding of system organization and dynamics that will allow formulation of general models of operation. These general models of operation should both guide and predict. Guide in the sense that as models are used, the formulation of system organization and dynamics will be refined or changed to better reflect reality. Predict in the sense that the model provides a method to test validity by predicting a future, which can be evaluated. The final issue is that even specified objectives, such as the restoration and maintenance of ecosystem integrity can be decomposed into critical attributes that must be evaluated both singly and in concert with all other attributes. This means that watershed management with an ecosystem emphasis is much more complicated than typically recognized. It means that a new paradigm, which starts from ecosystem needs, is needed in modern watershed management.

Acknowledgements

I would like to acknowledge the contributions of Steven Russel and Michael Schwar to the conceptual development of the Illinois River Ecosystem Restoration Program ecological integration process. This research supported in part from U. S. Environmental Protection Agency STAR research projects R 827451, Integrated Assessment of Climate Change Impact in the Mackinaw River Watershed, Illinois, and R 827148 Development of an Integrated Scientific and Technological Framework for Stream Naturalization.

References

Bledsoe, B. P. (2001). Relationships of stream responses to hydrologic changes. Urbonas, B (ed) Linking stormwater BMP Designs and Performance to Receiving Water Impacts Mitigation. ASCE/Engineering Foundation Conference, Snowmass, CO. August 19-24, 2001.

Borman, F. H., G. E. Likens, D. W. Fisher, and R. S. Pierce (1968). Nutrient loss accelerated by clear cutting of a forest ecosystem. Science 159:882-4.

Close, M. E. and R. J. Davies-Colley (1990). Baseflow water chemistry in New Zealand rivers 1. Influence of environmental factors. New Zealand J of Marine and Freshwater Res. 24:343-356.

Edwards, P. J. and J. D. Helvey (1991). Long-term ionic increases for a central Appalachian forested watershed. J. Environ. Qual 20:250-255.

EPA. (1993). Watershed Protection: Catalog of Federal Programs. EPA 841-B-93-002. Washington, DC: Office of Water, Assessment and Watershed Protection Division.

EPA. (1995). Watershed Protection: A Statewide Approach. EPA 841-R-95-004. Washington, DC: Office of Water, Assessment and Watershed Protection Division.

Goudie, A. (1990). The Human Impact on the Natural Environment (3rd Edition). MIT Press, Cambridge, MA.

Hall, A. D. and R. E. Fagan. (1956). Definition of a system. Gen Sys 1:18-29.

Herricks, E.E. (1996). Water Quality Issues in River Channel Restoration, River Channel Restoration, Brookes A.W. and D.J. Shields (eds.), John Wiley and Sons, N.Y., pp. 179-200.

Herricks, E. E. and I. Milne. (1998). A Framework for Assessing Time-Scale Effects of Wet Weather Discharges. Final Report of Project 92-BAR-1. Water Environment Research Foundation. Arlington, VA.

McDowell, W. H. and G. E. Likens (1988). Origin, composition, and flux of dissolved organic carbon in the Hubbard Brook valley. Ecological Monographs 58(3):177-195.

NRC (National Research Council). (1991). Opportunities in Applied Environmental Research and Development. National Academy Press, Washington D.C. 173 pp.

Poff. N. L. (1997). Landscape filters and species traits: towards mechanistic understanding and prediction in stream ecology. J. N. Am. Benthol. Soc. 16(2):391-409.

Schaeffer, D.J., E.E. Herricks, and H.W. Kerster. (1988). Ecosystem Health: I. Measuring Ecosystem Health. Environ. Management 12:4, p. 445-456.

Shaver, E., J. Maxted, G. Curtis, and D. Carter (1994). Watershed protection using an integrated approach. Stormwater NPDES Related Monitoring Needs. Torno, H. C. (ed), ASCE, New York, p 435-459.

Vannote, R. L. G. W. Minshall, K. W. Cummins, J. R. Sedell, and C. E. Cushing. (1980). The river continuum concept. Canadian Journal of Fisheries and Aquatic Sciences 37:130-137.

Vitousek, P. M., J. R. Gosz, C. C. Gruer, J. M. Melillo, W. A. Reiners, and R. L. Todd (1979). Nitrate losses from disturbed ecosystems. Science 204:469-73.

Webster, J. R. and B. C. Patten (1979). Effects of watershed perturbation on stream potassium and calcium dynamics. Ecological Monographs 49(1):51-72.

Effectiveness of Integrated Reservoir Watershed Conservation in Taiwan

Jing-San Hwang[1], Rong-Maw Shen[2] and Jau-Pau Wang[3]

Abstract

Water resources development, the status of reservoir management, and the management problems of reservoir watershed in Taiwan are investigated. The effectiveness of integrated reservoir watershed conservation is evaluated. Several improved measures are planned in the integrated reservoir watershed conservation project to achieve four main goals, including (1) the sustainable development of reservoirs, (2) the ecological environment of water resources, (3) the integrated control of disasters mitigation, and (4) community-oriented management.

Introduction

The distribution of rainfall is rather non-uniform in Taiwan. The loose topsoil of mountainside in the upstream watershed results in high erosion by the storm. Without effective management, the intensification of land use has resulted in serious reservoir sedimentation, that is, the functions of reservoirs has been depleted. In the other hand, the rock strata of upland watershed in Taiwan are sheared, fractured, weathered and rather weak. There are fewer suitable locations for constructing reservoirs. The cost of the reservoir construction has been increased quickly. Besides, reservoirs formed by dams block the rivers, and they may create huge impacts on human and ecological environment. Constructing reservoir can causes many problems Shich way not be resolved efficiently, such as compensation to local people and watershed management. To build a reservoir nowadays is hardly supported by the public because of environmental protection concerns.

---

1. The Acting Director General, Water Resources Bureau, Ministry of Economic Affairs, Taipei, Taiwan
2. The Deputy Division Chief, Water Resources Bureau, Ministry of Economic Affairs, Taipei, Taiwan
3. Engineer, Water Resources Bureau, Ministry of Economic Affairs, Taipei, Taiwan

Today, the most important concerns in Taiwan is to restore natural environment, implement soil conservation and conserve water resources. In addition, it is essential to encourage the public by recognizing and participating in the conservation work of water resources to preserve the functions in a reservoir. In this article, water resources development, the status of reservoir management, and the management problems of reservoir watershed in Taiwan are investigated. The effectiveness of integrated reservoir watershed conservation is evaluated. Several improved measures are planned in the integrated reservoir watershed conservation project to achieve four main goals, including (1) the sustainable development of reservoirs, (2) the ecological environment of water resources, (3) the integrated control of disasters mitigation, and (4) community-oriented management.

Water resources development in Taiwan

The total area of Taiwan, an island of the Republic of China, is 36,000 km^2, and the mountain terrain above 1000 meters in elevation occupies about 32% . Because of steep mountains and short rivers, the rainfall runs to the sea quickly after it falls to ground. As a result, it is hard to save and use water. There are narrow plains in the western part of Taiwan, which cover the area of 9000 km^2, a quarter of the total area. The western plains have high population density and important areas of the economic development in Taiwan.

Annual rainfall of Taiwan averages 2,510mm, about 2.6 times of the worldwide amount (970 mm/yr.). Conversion of rainfall to water yield is about 90.5 billion m^3 so that Taiwan is the area of abundant rainfall. However, owing to high population density, every person per year gains about 4,250 m^3 of the rainfall. It is about one sixth of worldwide rainfall average. Table 1 lists the worldwide comparison of annual and per capita rainfall average. Therefore, Taiwan is the area of low potential of water resources.

The distribution of rainfall in time and space is non-uniform on the Taiwan Island. About 80% of the rainfall occurs between May and October, and especially the most of rain falls in the typhoons event. There are over three hundred typhoons passing through the island for the last 100 years. If the rainfall from typhoon is not enough, it would cause water shortage. The duration from November to April is the low flow period, typically in the south Taiwan. Only 10% of the total precipitation fall during low water period. Thus, it is rather difficult in allocation and management of water resources.

The data collected from 1989 to 1998 showed that the total water usage in Taiwan was 18.1 billion m^3, including agricultural water usage of 13.7 billion m^3 which was about 75.81% of the total water usage, industrial water usage of 1.7 billion m^3 which was about 9.24 % of the total water usage. The main water source is surface water of 11.8 billion m^3 (including 7.6 billion m^3 of water directly diverted rivers, and 4.5 billion m^3 from the reservoirs), 65.26% of the total water usage. And the remaining 34.74% of the total water usage is groundwater about 6.3 billion m^3. Figure 1 shows the water resources and usage in Taiwan.

Table 1. Worldwide comparison of annual and per capita rainfall average

State	Population (10^5)	Area ($10^3 km^2$)	Annual rainfall average (mm/yr.)	Annual precipitation average ($10^8 m^3$/yr.)	Per capita rainfall average (m^3/yr. /person)
Worldwide	556,000	150,000	973	1,459,500	26,250
U. S.	21,361	9,363	700	71,159	33,313
France	5,291	551	750	4,133	7,811
Iran	3,301	1,648	250	4,120	12,481
China	120,000	9,597	660	63,340	5,278
Japan	11,193	377	1,788	6,749	6,030
Taiwan	2,130	36	2,515	905	4,250

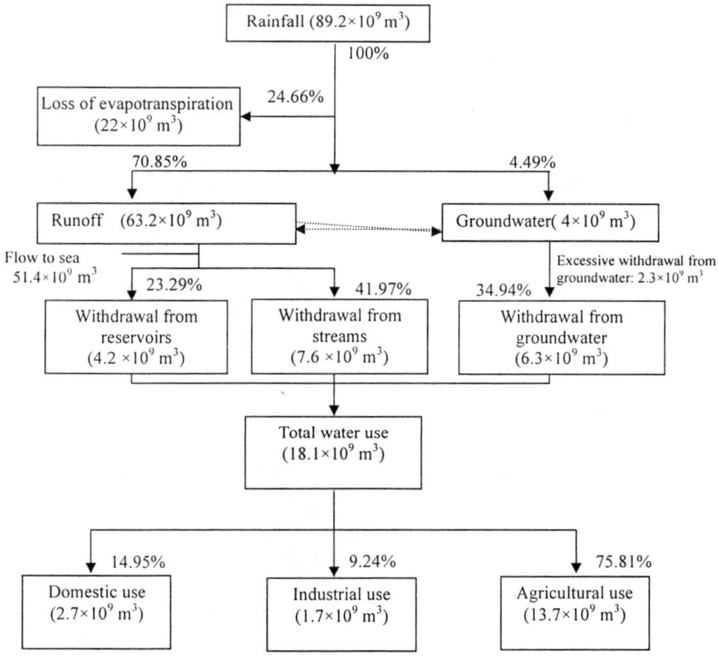

Fig. 1. Water resources and usage in Taiwan (1997)

The Status of the reservoir management in Taiwan

Due to the temporally and spatially non-uniform distribution of rainfall, a reservoir provides the capacity to store the surface water for water supply in Taiwan. There are about 88 reservoirs in Taiwan area including 40 major reservoirs. The area of the major reservoir watersheds is 5,200 km^2, about one seventh of the total area of Taiwan. The total capacity of major reservoirs is 2.05 billion m^3. Most of the reservoirs were built on the main streams in Taiwan.

The sites of reservoirs in Taiwan sites are shown in Fig .2. According to various functions, the reservoirs may be classified as: the single-purpose reservoir such as the Xishi, Xinshan, and Nanhua Reservoir for domestic water supply, and the multi-purpose reservoir such as the Feitsui, Shimen, Tehchi, and Zengwen Reservoir for irrigation, domestic water supply, hydraulic power, or flood control.

According to the result of survey in 1999 (see Table 2), there are 40 reservoirs with operation records, and the total capacity changes from 2.736 billion m^3 to 2.05 billion m^3. Annual sediment yield is over 15 million m^3, and the reservoir sedimentation problems in the southern region is more serious. Presently, reservoir sedimentation rate of 50% or more happens in Guguan , Agongdian, Luliaoxi, Jianshanpi, Wushantou, Hutoupi, and Yanshuipi reservoirs. The sedimentation problems of reservoir are usually serious. It may be caused by the storm during typhoon season, steep topography, weak geological formations, frequent earthquake, and human destruction, etc. In the aspect of active storage, the Zengwen Reservoir has its active storage about 5.7×10^8 m^3; the Feitsui Reservoir has about 3.4×10^8 m^3; and the Shimen Reservoir has about $2.3\times10^8 m^3$. And in the aspect of the height of the dam, the highest dam is Tehchi Reservoir of 180 m. The height of the Shimen Reservoir is 133.1 m, and the height of the Zengwen Reservoir is 133m.

Table 2. The data on sedimentation of reservoirs in Taiwan

Reservoir name	catchment area (km^2)	initial total capacity (million m^3)	Active storage 1999 (million m^3)	sedimentation amount (million m^3)		Sedimentation rate (%)
				total	annual average	
Hsinshan	1.60	6.00	5.31	0.44	0.266	7.38
Hsishih	6.50	0.65	0.44	0.20	0.003	30.77
Feitsui	303.00	406.00	339.70	17.80	1.155	4.038
Chihtan	717.00	4.20	3.63	0.57	0.027	13.57
Shihmen	763.40	309.12	232.77	58.81	1.657	19.02
Paoshan	3.20	5.47	4.43	0.92	0.064	16.87
Tapu	104.00	9.40	5.42	3.90	0.099	41.44
Yunghoshan	4.80	29.58	27.25	1.36	0.090	4.60
Mingteh	61.08	17.70	13.91	3.60	0.122	20.32
Tehchi	592.00	255.41	170.20	25.00	0.962	9.79
Kukuan	707.75	17.10	6.85	10.10	0.265	59.06
Shihkang	1,061.00	2.70	-	-	-	-
Wusheh	219.00	148.60	91.42	57.18	1.375	38.48
Tousheh	0.50	0.30	0.23	0.02	0.001	6.67
Liyutan	53.45	126.07	119.88	0.64	0.090	0.51
Jenyitan	3.66	29.11	25.85	2.00	0.160	6.85
Lantan	30.00	9.80	8.92	0.32	0.012	3.23
Luliao Creek	7.50	3.78	1.18	2.60	0.043	68.72
Paiho	26.55	25.09	13.26	11.83	0.343	47.15
Teyuanpei	32.11	3.85	1.74	2.07	0.480	53.77
Chienshanpei	10.28	8.11	3.16	5.05	0.830	62.27
Tsengwen	481.00	712.70	568.97	105.46	3.955	14.80
Nanhua	104.00	154.41	134.55	16.41	2.698	10.63
Wushantou	60.00	171.19	83.75	87.44	1.257	51.08
Chingmien	2.73	0.15	0.99	0.13	0.007	11.39
Yenshuipei	5.73	0.76	0.25	0.51	0.012	67.11
Hutoupei	7.15	1.36	0.80	0.52	0.007	38.24
Akungtien	31.87	35.49	21.88	13.61	0.302	38.35
Chengching Lake	2.88	5.00	4.39	0.50	0.009	10.00
Fengshan	2.75	9.20	7.80	1.05	0.067	11.44
Mudan	69.20	33.95	29.08	4.55	1.011	13.40
Lungluantan	4.37	3.76	3.52	0.09	0.002	2.26
Chengkung	1.36	1.08	1.05	0.10	0.00	1.11
Tungwei	1.30	0.19	0.16	0.30	0.001	14.21
Hsingjen	2.27	0.68	0.69	-0.06	-0.003	-8.68
Hsian	0.82	0.24	0.24	0.00	0.000	0.00
Hsiaochih	1.05	0.19	0.04	0.15	0.016	80.20
Chimei	1.14	0.23	0.23	0.00	0.000	0.00

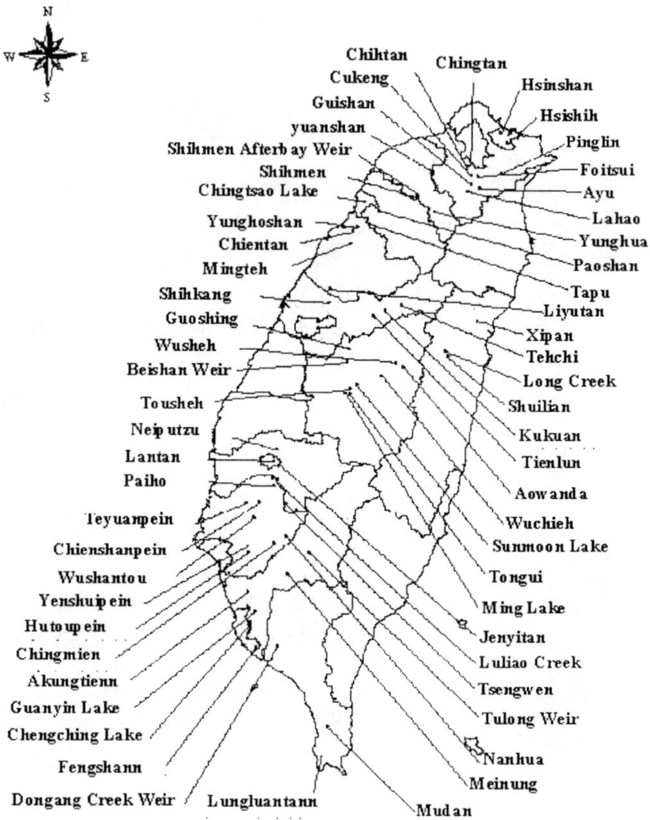

Fig.2 Reservoir sites in Taiwan

According to the statistic data about the water quality in reservoirs from Environmental Protection Administration (table 3), the quality of reservoirs in 1999 was much better in 1998. The water quality of six reservoirs had been improved from high eutrophication to the general condition of eutrophication. It showed that there was some achievement contributed to reservoir watershed management. But the eutrophication condition of water quality of major reservoirs was still serious in Taiwan. Table 3 indicates that reservoirs with TSI reaching 50 include Liyutai, Chengqing lake, Jingmian, Fengshan, Baihe, Lantan, Renyitan、Agongdian, etc. In general, water quality has become eutrophicated, the polluted water could not be treated economically and effectively by traditional equipment of the drinking treatment plants. And it could not serve as raw water of drinking use. For shortage of water resources in Taiwan, the eutrophication of water quality of reservoirs was negative for water resources management.

Table 3. Carlson's Trophic Status Index （TSI） for major reservoirs in Taiwan

Reservoir name	1993	1994	1995	1996	1997	1998	1999
Xinshan	51.74	49.43	43.50	51.90	48.98	--	52.57
Xishi	47.92	47.36	--	--	--	--	--
Feitsui	--	40.85	36.00	45.81	45.09	44.61	45.16
Shimen	45.56	55.76	42.75	51.19	49.34	50.20	42.98
Baoshan	50.30	47.55	48.00	50.48	50.16	54.85	49.27
Tapu	62.86	--	--	--	--	--	--
Xingren	--	89.20	--	--	--	--	--
Dongwei	--	78.45	--	--	--	--	--
Yongheshan	51.33	47.42	43.25	49.47	51.67	53.95	47.60
Mingde	53.05	52.92	45.75	47.22	51.81	52.50	48.03
Tehchi	--	47.90	40.75	52.52	54.85	56.40	38.77
Wushe	45.44	40.42	41.00	42.33	50.87	46.12	42.82
Liyutan	--	--	49.25	47.73	54.34	53.94	51.13
Sunmoon lake	36.01	34.99	35.50	38.32	42.56	47.15	42.96
Renyitan	56.91	59.62	51.50	47.51	55.58	57.72	50.63
Lantan	53.74	50.85	45.25	47.47	57.41	--	53.93
Baihe	51.36	56.20	49.00	55.94	55.25	56.12	53.61
Zengwen	51.00	49.81	40.75	47.17	51.57	50.49	44.49
Wushantao	45.06	50.88	44.50	54.92	47.58	47.29	43.02
Nanhua	--	--	46.75	50.68	77.14	51.23	44.08
Jingmian	59.01	56.11	46.25	51.19	59.66	51.55	55.20
Agongdian	75.08	80.79	69.50	84.37	79.14	--	--
Chengqinhu	72.74	67.94	65.50	72.31	73.85	69.65	66.57
Fengshan	77.53	74.56	75.00	79.06	55.20	78.50	74.73

Source： Environmental Protection Administration

The management problems of reservoir watersheds

Because of dense population, rapid development of economics, all kinds of human activities frequently and land scarcity of plains, people have great expectancy and requests for development of land of mountains and hills. This is not very favorable to strict implementation of the integration conservation of reservoir watershed and regulation system. The common destruction behaviors in reservoir watersheds include as follows:

(1) Many forested slopes have been cleared and cultivated illegally. The plants of high economic price such as betel palms, tea trees, vegetables, and fruit trees have replaced the wood. This influences water conservation, and causes serious soil erosion and debris flow. Owing to soil erosion of upland watersheds, the capacity of reservoirs reduces rapidly and it shortens the life of reservoir.

(2) All kinds of activities such as roads construction, foundations building and mining, etc., often change the landscape. It affects natural system of drainage and stabilization of mountain and hill slopes, and causes soil erosion easily. When storm comes, the peak of runoff will increase.

(3) A large amount of fertilization and pesticide for vegetables, fruit trees, and tea trees may affect the water quality of reservoirs.

(4) Increasing recreation activities and tourists' behaviors such as camp, barbecue, and excretion, have imposed great threats on the water quality of reservoirs. And it produces eutrophication of water quality of reservoir.

Besides the above destruction behaviors, there are some restrictions for reservoir watersheds like the protection area of water quality and water quantity of water source, which result in strong opposition of local public. In the present, the most important thing is to integrate and coordinate related regulations and government organizations. And other problems include:

(1) A number of agencies involved with watershed management, each with its own respective mission, role, priority and authorized regulations is problematic.

(2) Insufficient management members of agencies.

(3) The system of feedback in reservoir watersheds is not intact.

(4) Treatment and reuse of reservoir sedimentation.

(5) Natural disasters of watershed keep happening.

(6) Traditional engineering measures may be detrimental to ecological environment of watershed.

The effectiveness of reservoir watershed conservation management

The large reservoirs in Taiwan like Feitsui, Shimen, Tehchi and Zengwenall reservoirs have their own administrative organizations which set up watershed management projects, and have budget each year to perform conservation works including forest management, water and soil conservation of agricultural land, landslide treatment, eroded ditch management, sediment control engineering, water quality monitoring and management, education, etc. These projects have produced some major achievements. From 1997, "the integrated project of conservation in medium and small reservoir watersheds" is undertaken and get elementary effectiveness.

1、Large reservoir watersheds

(1) Administrative organization of Feitsui Reservoir prepares budget each year to carry

out water and soil conservation works. Budget for management in 1999 was about 6.5 billion dollars (table 4), conservation works include landslide management, reforestation, slopeland vegetation, Shoot crete, retaining wall, streambed stability work, etc. To prevent flushing downstream by flood discharge of reservoir, the regulation works along the river reach near the dam have been done. Activity of planting trees was held once year for coserving water resources and invoking the public to love forest.

(2) Administrative organization of Shimen Reservoir is carrying out the second stage regulation project (1992~2001). The regulation results contained field treatment of agricultural land of 367ha, mulching of 16ha, system of drainage of 4936ha, 10 sediment storage dams, gully control of 1146m, road sideslopes vegetation of 17.9ha, 8 road sideslopes controls, 75 landslide controls, 6 streams controls, catchment afforestation of 315 ha, etc. The works completed for the past years are shown in Table 5. The budget for the third stage regulation project (2004-2015) will be about 3.675 billion dollars.

(3) Administrative organization of D techi Reservoir is carrying out the fourth stage regulation project (1998-2002), and the project works of 2001-2002 are shown in Table 6. The budget contained 161 million dollars for forest management, 172 million dollars for water and soil conservation, 133 million dollars for sediment control engineering, 82 million dollars for reservoir protection zone, 71 million dollars for reducing pesticide and public education, 39 million dollars for water quality monitoring and management, and 683 million dollars for environmental conservation and landslide control engineering, etc.

(4) Administrative organization of Zengwen Reservoir had implemented three stages of management project form 1967 to 1991. The total cost spent was about 870 million dollars. The fourth stage project since 1992 has been implemented and continued for 10 years. The major wroks include river regulation, landslide control, water and soil conservation of road, field treatment of agricultural land, and forestry management. Total expenditure was about 850 million dollars (Table 7).

2、Medium and small reservoir watersheds

According to " The conclusive and suggestive action plans of National Water Conservancy Conference in 1994", the integrated conservation project of medium and small reservoirs was set up. It was an integrated project of 5-year period (1997~2001), and its contents were based on the reports of regulation planning of 23 medium and small reservoir catchments in Taiwan. Its major works contained sediment control engineering, afforestation breeding, water and soil conservation of agricultural land and road, reservoir sedimentation dredging, reservoir protective zone, landslide control, and gully control (Table 8). Following was the effectiveness of the integrated project:

(1) Stabile facilities of mountain and hill slopes

There were retaining wall of 15,504m, sideslopes revetment of 1,440m, and 4 retaining walls constructed. These could stabilize sideslopes, prevent landslides, and protect houses and people lived in downslopes indirectly. The facilities may prevent landslides of 438,120 m^3 soil.

(2) Drainage facilities

There were drainage ditch of 2360m, L-shaped ditch of 1068m, and 11 culverts built. These could efficiently divert runoff to safe area downstream, prevent soil erosion, and decrease the damage of destruction to the minimum. These utilities may protect the area including 9,285 persons, 782 factories and houses.

(3) Road maintenance

There were roads of 15,775m repaired and maintained. These could avoid roads construction to increase government finance or affect social stability and economic development.

(4) Reservoir dredging

The total amount of sediment removal is about 702,786 m^3, and approximately equal to the capacity of XiShi Reservoir. Dredging did not only increase reservoir capacity but also improved water quality, and elongated life-time of reservoir. In the consideration of environmental conservation and cost of reservoir construction, it was a good engineering measure for increasing available water resources.

(5) Stream regulation engineering

These engineering structures included bank revetment of 24,527m, 105 stream-bed stability works, 19 submerged dams, and cement work of 8,058m. They could stablelize riverbank, control current, prevent transverse and vertical flow erosion and landslide, protect side slope of riverbank and riverside public facilities. They could protect riverside of 36,345m and prevent erosion sedimentation of 407,320m^3.

(6) Sediment storage facilities

These facilities contained 11 main check dams. They could store sediment from river, regulate sediment transporting, retard the slope of riverbed, control the current, prevent landslide, restrain debris flow, and mitigate disasters. They may avoid upstream sediment of 42,400m^3 into reservoirs.

(7) Across drainage facilities

There were 7 across drainage facilities constructed. These could intercept upstream runoff to safe sites, and have effectiveness of protecting agricultural and wood lands of 2,787ha.

(8) Energy dissipated facilities

These facilities contained 14 drop structures and energy dissipated pools, which could decrease flow velocity and have the effectiveness of protecting 43 bridges and one school.

(9) The vegetative cover

These works contained landscape vegetation of 90,713m^2, afforesting and breeding of 67ha, and reservoir protective zone of 1,971m. They could help to conserve water, avoid land erosion and landslide, mitigate flood damage, create clear and beautiful environment, and have effectiveness of decreasing soil erosion 38,500m^3 each hectare every year.

3、The improvement of reservoir sedimentation and erosion of reservoir watersheds

After the long-term conservation management, it is very significant to reduce the soil erosion and sedimentation of large reservoirs for the recent years. Table 9 shows the annual sedimentation in Feitsui, Shimen, Zengwen and Tehchi are reservoirs. The amount of sedimentation for the last 2 years is far less than that of the initial planning. It is not only to elongate the lifetime of reservoir but also to slow down sedimentation of reservoir.

In the part of medium and small reservoirs, good results on improvement of reservoir sedimentation can be shown from Fig.3 to Fig.6. These figures indicate the difference between the cumulative sedimentation with or without conservation project in Mingde, Tapu, Baihe and Akongdian reservoirs, and the sedimentation of reservoirs with conservation project has been substantially improved.

Table 4 The budget and regulation facilities of Feitsui Reservoir in 1999

Year	Landslide Mamage -ment (1000NT)	reforestation (1000NT)	Shoot crete (m^2)	Retaining wall (m)	Sediment storage dam (numbers)	Drainage ditch (m)	Replant area (ha)	Breeding area (ha)
1999	72,584	62,810	4,301	887	1	489	353	1,978
1964 ~1998	493,358	156,410	9,817	14,687	13	5,715	459	2,815
Total	565,942	229,220	14,118	15,574	14	6204	812	4,793

Table 5 Facilities of regulation in Shihmen Reservoir

Item / year	Field treatment (ha)	Mulching (ha)	Drain system (ha)	Check dam (No.)	Gully control (m)	Road Sideslopes control (No)	Landslide control (No.)	Catchment of afforestation (ha)	River Dredging	Extension education (class)	Aerial survey (set)
1964 ~1991	1,910	2,800	53,600	105	26,500	52	204	13,612	21	786	3
1992 ~1997	376	16	4,963	10	1,146	8	75	315	6	22	1

Table 6 Budget of regulation in the Tehchi Reservoir, 2001-2002 (unit: $1000NT)

Item / year	Forest manage ment	Soil and water conservation	Road engineering	fertilizers decreasing enforcement	Water quality monitoring and management	Ecology conservation education	Sediment control	Reservoir protective zone	Total
2001	38,314	53,119	15,965	9,858	10,913	1,700	17,050	14,735	161,654
2002	18,758	36,085	17,965	9,858	10,835	1,275	5,900	7,045	107,721

Table 7 The contents of the fourth stage conservation project for Tseng-Wen reservoir

Item	River regulation engineering	Landslide control	Water and soil conservation of road	Field treatment of agricultural land	Landslope conservation and forest management	Others	Total
Content	15check dams	41	31	500ha	—	—	—
Budget (1000NT)	341,501	36,085	171,419	10,000	47,020	73,393	850,223

Table 8 The contents of the medium and small reservoirs conservation project (unit:numbers)

Item	1997	1998	1999	2000	total	%
(1) Sediment dredging	0	2	2	10	14	3.97
(2) Reservoir protective zone	4	3	3	2	12	3.40
(3)Vegetation recovery	0	3	11	10	24	6.80
(4)Landslide control	11	17	24	23	75	21.25
(5)Gully control	18	29	33	55	135	38.24
(6)Water and soil conservation of agricultural and road	12	8	11	25	56	15.86
(7) Water quality protection	0	0	0	0	0	0.00
(8)Management and new disaster treatment	6	2	2	3	13	3.68
(9)Others	5	9	7	3	24	6.80
Total	56	73	93	131	353	100

Table 9 The statistics of Large Reservoir sedimentation (unit : $10000m^3$)

Reservoir name	Total storage	Annual sedimentation of plan	Annual sedimentation average	Annual sedimentation (the recent year)
Feitsui	40,600	113.6	101.0	84.0 (2000)
Shihmen	30,912	80.0	86.4	27.4 (2000)
Tsengwen	712,700	561.0	379.9	92.0 (2000)
Tehchi	26,220	214	98.5	47.5 (2001)

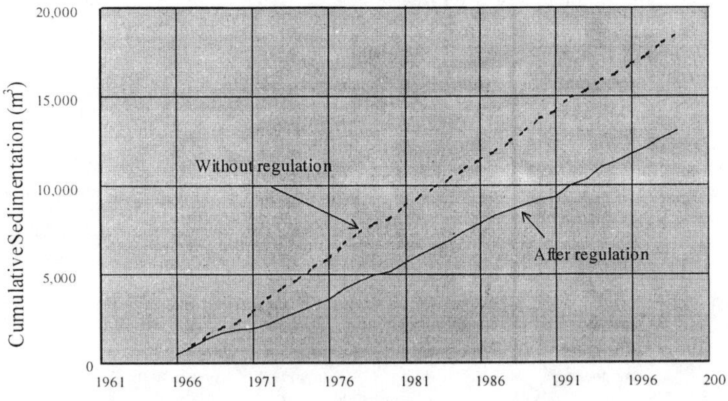

Fig. 3 The effect of watershed regulation on sedimentation reduction in Baihe reservoir.

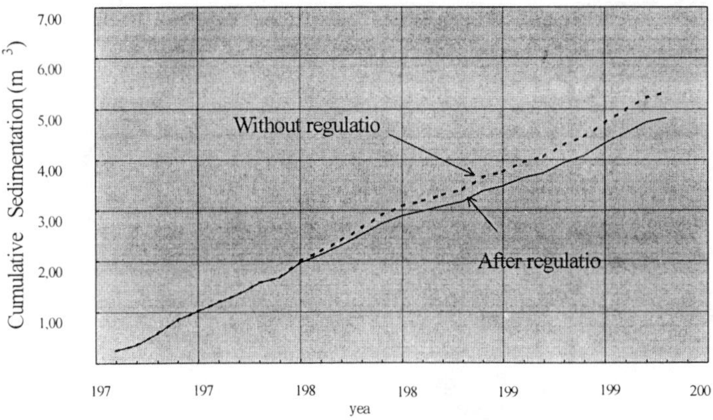

Fig.4 The effect of watershed regulation on sedimentation reduction in Mingde reservoir

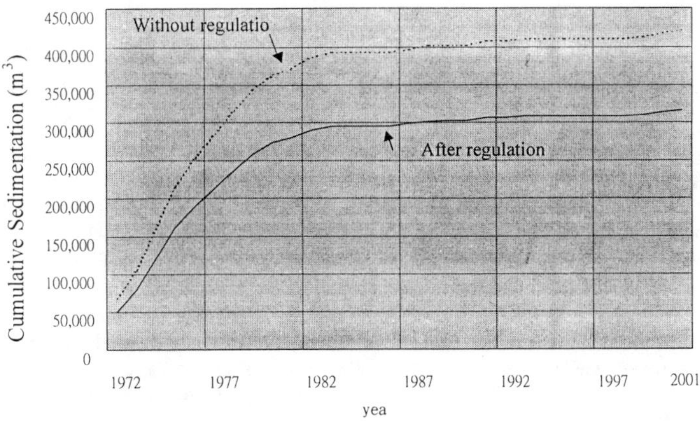

Fig. 5 The effect of watershed regulation on sedimentation reduction in Dapu reservoir

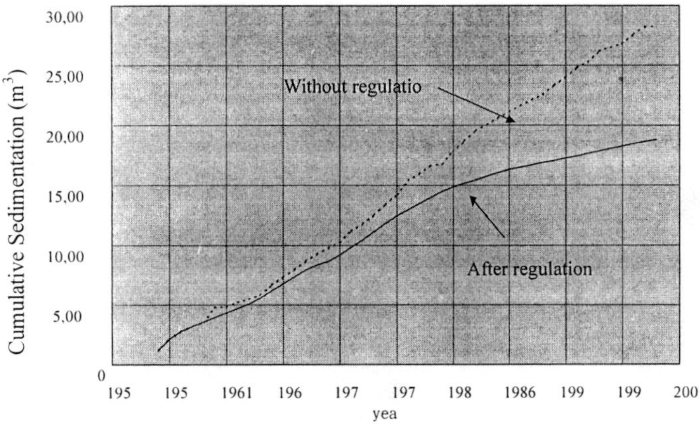

Fig.6 The effect of watershed regulation on sedimentation reduction in Agongdian reservoir

Improved measures of reservoir watershed conservation management

After evaluating the effectiveness of reservoir watershed conservation management, some improved measures are provided as follows:

1. Nowadays, management projects of large reservoir watersheds often lacks sufficient budget to carry out. Conservation works are also constructed by traditional engineering measures, and the project without considering water-intimate, ecological engineering, and community construction. Therefore, the effectiveness of regulation and conservation is not significant. A medium / long-term integrated project must be proposed, and budget approved should be increased to invest on the conservation works.
2. The total engineering facilities constructed in medium and small reservoir watersheds conservation project only reached 29.12% of the overall works. It showed that the real budget approved was far less than that of the project , and the goals could not be achieved. Therefore, the total budget approved every year should be increased for enhancing effectiveness of watershed regulation and conservation.
3. The achievable rate of conservation works such as reservoir dredging, vegetation covering, gully control, water and soil conservation of agricultural land, and improved facilities of water quality was low for integrated conservation project of medium and small reservoirs. But now the reservoirs in Taiwan face the problems of the capacity reduction and water quality deterioration. Therefore, the above five conservation works will be focused in the next project.

4. In order to change negative impression brought by reservoir operation to the public, efforts must be concentrated on the works like water-intimate and biological engineering measures, conservation and education, which are expected to establish a water-intimate, harmonic reservoir community and the biological environment, and through public participation and support for ensuring effective implementation of proper land use planning and regulation.
5. The sedimentation of reservoirs in Taiwan area is serious, and there are 7 reservoirs with sedimentation rate reaching 50% , especially in the southern regions. To ensure the capacity of reservoirs, the works of reservoir dredging, study on treatment and reuse of reservoir sedimentation and enactment of relevant regulations should be made.
6. Earthquakes, frequent seismic shakings, heavy typhoon rainfalls, and weak geological formations cause broken land and loose topsoil of reservoir watersheds in the recent years. Considerable efforts have been made for the protection of land, conservation of water resources and control of disasters migition, and the conservation works should be extended to all the reservoir watersheds, Therefore, it is extremely important to propose an integrated conservation project of reservoir watersheds to achieve the four targets of sustainable management of water resources ： a) the sustainable development of reservoirs ； b) the ecological environment of water resources ； c) the integrated control of disasters mitigation ； and d) the community-oriented of management.

Conclusion

"Second World Water Forum"-Ministerial Declaration of the Hague on Water Security in the 21th century stated that "the government have to supply clean, sufficient, and safe water". However, there are not abundant water resources available in Taiwan, and the distribution of the rainfall in time and space was so uneven that reservoirs were very important to regulate water resources. With the rapid development of economics, urbanization and intensive industrialization in the past years, the water demand increase rapidly in Taiwan. But it also promotes development and destruction in upstream watershed. As a result, water was polluted, and it affected the available water resources. Additionally there was more and more obstruction in exploiting new water resources. How to conserve the limited water resources of reservoirs were the most important study.

Building reservoirs may not be the best choice to develop new water resources for the present biological environment in Taiwan. To ensure the limited water amounts of reservoirs to play an important role in regulation and allocation of water resources, a medium/long-term integrated conservation project of reservoir watersheds has to be proposed as soon as possible, which include conservation and renewal of reservoir, watershed management, removal of reservoir sediment, improvement of water quality, and renewal of reservoir facilities, etc. Efforts must be concentrated on developing appropriate institutions and policies, fostering watershed land stewardship, controlling human behaviors on watershed, reservoir dredging, forest management, bioengeering measures, combination of structure and non-structure conservation measures, and a properly implemented public information, education and involvement program to achieve the goal of sustainable management of reservoir watershed land and water resources.

Reference*

1. Chia-Nan Irrigation Association, 1988, "The Report of Soil and Water Conservation Measures Analyses in Wushantao Reservoir Watershed."

2. Tsen Wen Reservoir Administration Bureau, 1989, "Tsen Wen Reservoir Management Evaluation Studies."

3. Shao, W.K., 1993, "Development of Water Resources and Reservoirs in Taiwan,"Journal of Civil and Hydraulic Engineering, 19(4) : 43-57.

4. Chang,S.L, 1996, "Future Role of the Watershed Conservation and Disaster Prevention in Taiwan," Proc. of the Hillslope disaster prevention Symposium.

5. Department Of Water Resources , 1996, "Integrated Conservation Program of Small/Medium Reservoirs."

6. Water Resources Bureau, 1998, "Reservoirs."

7. Water Resources Bureau, 1999, "An Introduction to River Ecological Environment Protection."

8. Wu,R.S, S.C. Lin and M.H.Wang, 1999, "A Study of the Reservoir Planing and Management for Public Water Supply Goal," Civil and Hydraulic Engineering, 26(2) : 78-93.

9. Shiang-Kween, Hsu, 2000, "An Introduction to Water Resource Sustainable Development ,"

10. Water Resources Bureau, 2000, "Study on Small and Medium Reservoirs Management Planning and Evaluation."

11. Water Conservancy Agency, 2000, "Reservoir & Dam Project Data."

12. Cheng-Daw Hsieh,2000, "The Regulation and Conservation of the Reservoirs Watersheds in Taiwan ," Proc. of the Watershed Protection Symposium.

＊All the papers of referen

Evaluation of Urban Runoff Control Programs

Michael A. Ports, P.E., P.H., Fellow ASCE[1]

Introduction

Although the overall quality of U. S. waters has improved measurably since the passage and implementation of the Clean Water Act of 1972, a significant number of water bodies continue to suffer from poor water quality. Over the years, most point sources of pollution from industries and municipal treatment plants have been brought under control. However, continuing water pollution suggests that other sources, including urban runoff, contribute to water quality problems. Thus, the Environmental Protection Agency (EPA) now classifies urban runoff as a significant cause of water quality impairment.

The Water Quality Act of 1987 requires EPA to regulate urban runoff that reaches municipal sewer systems as point sources of pollution. Under the authority of the act, EPA requires certain local governments to take measures to control urban storm water runoff. Concerned about water quality degradation in urban areas, Congress asked the General Accounting Office (GAO) to report on the amount of runoff from urban areas, particularly from roads and other impervious surfaces, and its effects on water quality. Congress further asked for an overall evaluation of the urban runoff control programs that federal regulations require of local governments, including their cost and effectiveness.

During late 2000 and early 2001, the GAO reviewed numerous reports from federal and other sources, interviewed experts in the field, and visited five urban areas and the organizations affiliated with those watersheds. The five urban areas include Baltimore, Maryland, Boston, Massachusetts, Los Angeles County, California, Milwaukee, Wisconsin, and Worcester, Massachusetts. The results from the GAO evaluation are summarized in a report published in June 2001 and titled, "WATER QUALITY, Better Data and Evaluation of Urban Runoff Programs Needed to Assess Effectiveness."

---

[1] Director, Environmental Engineering Services, HNTB Corporation, 715 Kirk Drive, Kansas City, Missouri 64105. (816) 472-1201. Mports@hntb.com

Background

Nonpoint source pollution results from water running over land surfaces and into bodies of water. Significant sources of nonpoint pollution may include urban land uses, agricultural fields and facilities, forestry activities, and mining operations. However, in urban and suburban areas, runoff generally enters a municipal sewer system that can be regulated as a point source of water pollution. In order to regulate urban storm water runoff, EPA published regulations in 1990 that established the National Pollutant Discharge Elimination System (NPDES) Storm Water Program and specified permit application requirements. The purpose of the program is to preserve, protect, and improve water quality by controlling the volume of runoff from paved surfaces and by reducing the level of runoff pollutants to the maximum extent practicable using best management practices (BMPs).

EPA provides federal funds and technical assistance to states for their own nonpoint source pollution management programs. At present, EPA manages the NPDES Storm Water Programs for six states, namely Alaska, Arizona, Idaho, Massachusetts, New Hampshire, and New Mexico. Thus, forty-four states manage their own programs under authority delegated by EPA. The program requires implementation in two primary phases. Phase I requires all municipalities with a population of 100,000 or more to obtain a permit for their municipal separate storm sewer system (MS4). In addition, storm water discharges from sites with industrial activities and construction sites that disturb more than five acres of land must obtain permits. Other municipalities and industrial activities may be brought into the program if deemed appropriate by the permitting authority.

In 1990, the regulations specifically identified 220 municipalities that were required to apply for a Phase I permit. According to EPA, 256 Phase I permits were issued and 17 more remained under review as of April 2001. Because some of the permit applications covered more than one municipality, the initial Phase I permits cover approximately one thousand medium and large cities nationwide.

In December 1999, the final rule for Phase II was issued. The rule extends the requirement for obtaining permits to all MS4 operators in urban areas not already covered under Phase I. Urban areas, as defined by the Census Bureau, have a residential population of at least 50,000 and an overall population density of at least 1,000 persons per square mile. The rule also extends permit coverage to construction sites that disturb at least one acre of land. EPA anticipates that as many as 5,000 municipalities may be required to submit permit applications under Phase II. The deadline for obtaining Phase II permits is March 10, 2003.

The Total Maximum Daily Load (TMDL) Program is intended to address those bodies of water that do not meet water quality standards because of pollutant loads from both point and nonpoint sources. It remains unclear just how and when the TMDL program will affect NPDES storm water permits. The program requires each state to allocate the pollution load, or TMDL, among the various point and nonpoint sources that flow into the subject water body and to take steps to ensure that no single source exceeds its assigned load. Congress has stayed the implementation of the TMDL rules. But, at some point, it will be possible that establishing a

TMDL for a particular water body will result in the application of specific numerical effluent limits for particular storm water outfalls.

Urbanization Causes Increased Storm Water Runoff

Since World War II, urban runoff has increased dramatically across the United States. That increase is related directly to the growth in the amount of impervious surfaces from residential developments, roads, highways, and other developments. The loss of natural vegetation, destruction of wetlands, and construction of storm sewer systems increases both the volume and velocity of the runoff. Numerous studies have documented that urban runoff and the pollutants it carries negatively impact water quality, aquatic life, and public health.

From 1945 through 1997, the area of land devoted to urban use in the contiguous United States increased by almost 327 percent, from 15 million acres to 64 million acres. More recently, from 1992 through 1997, the annual rate of new urban development averaged one million acres primarily from forest, pasture, and range. The amount of urbanized land in Baltimore, Maryland and Washington, DC grew by more than 170 percent from 1960 to 1990.

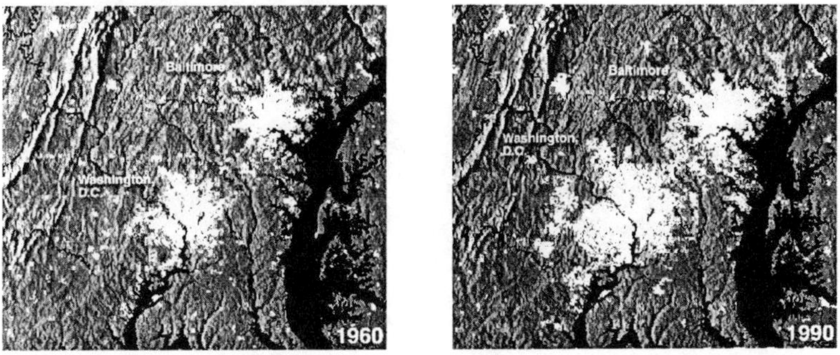

Increase in Urbanized Land, 1960-1990

The increase in paved surfaces has come from more than just new urban and suburban development. Our culture and economy have spurred the steady rise in the use of roads and highways. The automobile remains the primary mode of daily transportation for most Americans. Trucks carry three fourths by value of all goods shipped. From 1945 through 1997, paved road mileage in the country increased by 278 percent. In 1945, just 19 percent of public roads were paved. However, in 1997, that amount increased to 61 percent. One study suggests that motor vehicle infrastructure, including roads and parking lots, accounts for nearly half of all land area in our cities.

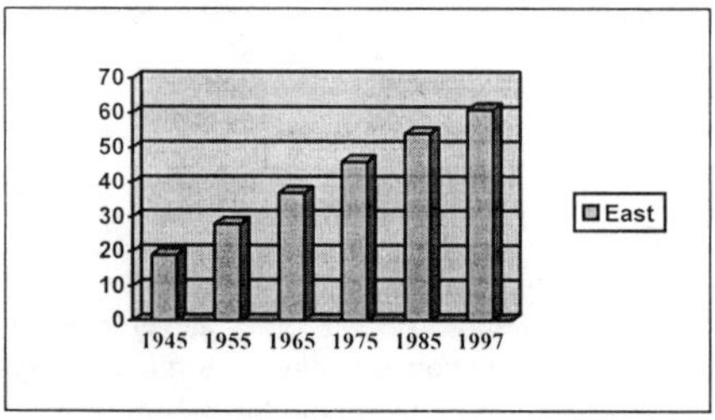

Percentage of Paved Public Roads, 1945-1997

As impervious surfaces increase, less storm water infiltrates into the soil. In general, paved areas allow only a small amount of initial infiltration compared to unpaved areas. EPA estimates that only ten percent of total precipitation runs off the land into nearby creeks, rivers, or lakes, when natural ground covers are present over an entire site. However, when a site consists mostly of impervious surfaces, as much as 55 percent or more of the precipitation runs off.

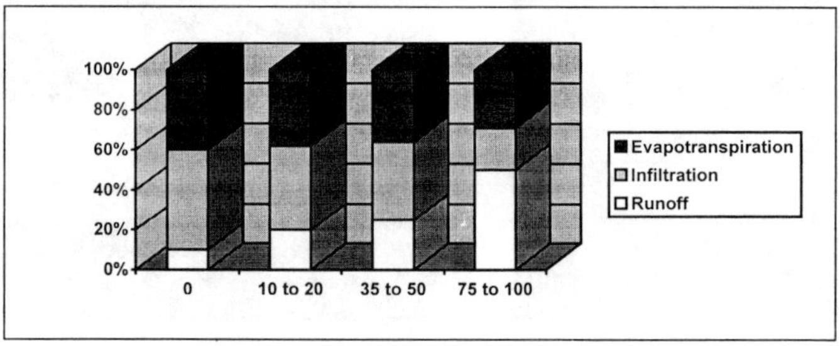

Evapotranspiration, Infiltration, and Runoff vs. Imperviousness

Urban Development Impacts on Water Quality

EPA has identified urban storm water runoff as one of the leading sources of pollution to rivers, streams, lakes, and estuaries. The Clean Water Act requires states to report on the quality of their waters every two years. The 1998 *National Water Quality Inventory Report to Congress* showed that 35 percent of assessed river and stream miles, 45 percent of assessed lake acres, and

45 percent of assessed estuarine acres were impaired in terms of ability to support aquatic life, swimming, and fish consumption.

As the percentage of impervious cover increases in a watershed, biodiversity declines. Recent research indicates that, in general, watersheds with less than ten percent imperviousness exhibit high diversity of fish and other aquatic life and high water quality. However, watersheds with as little as 25 percent imperviousness generally are geomorphically unstable, exhibit poor diversity of fish and aquatic life, and have poor water quality.

Damage Caused by Urban Runoff

Numerous studies associate urban runoff with fundamental impacts on receiving waters. Urban runoff carries more sediment and other contaminants that harm aquatic plants and animals. Urban runoff also may warm receiving waters. Less oxygen then is available for aquatic organisms. These and other factors, when combined, serve to degrade aquatic habitat, commonly reducing biodiversity and increasing invasive species. Several efforts have attempted to characterize the chemicals and other constituents in urban runoff. Most notably, EPA conducted the Nationwide Urban Runoff Program (NURP) between 1978 and 1983. The principal contaminants found in urban runoff include nutrients, solids, pathogens, metals, hydrocarbons, organics, salt, and trash.

Local Governments Take Action

In order to comply with the Phase I requirements, permitted municipalities must create and implement comprehensive storm water management programs. The three primary elements of the programs include efforts to characterize storm water runoff, BMPs aimed at reducing pollutants to the maximum extent practicable, and reporting program activities and monitoring results. More than one thousand cities are undertaking these efforts, yet only limited information

exists on the overall costs of managing urban runoff and the effectiveness of the programs. EPA's attempts to estimate costs are incomplete or out of date. The approaches used to estimate costs by the permitted municipalities are vary considerably. Both EPA and the delegated states believe that the program will be effective in improving water quality. But, no systematic effort to evaluate the program has been attempted. Without such an effort, EPA cannot tell what effect the program has on water quality nationally.
GAO found that the five local governments reviewed all were in the process of implementing their storm water management programs. First, all five were monitoring for a variety of purposes, including runoff characterization from different land uses in order to target their BMPs, testing the effectiveness of BMPs, or establishing baselines for their storm water quality evaluations. Second, all five generally were following a low cost, common sense approach to BMP implementation. While some structural BMPs, such as detention, filters, and buffers, are essential, more emphasis is placed on non-structural BMPs. Most permitted municipalities rely more heavily on non-structural measures, including

- Good housekeeping techniques, such as oil collection and recycling, spill response, household and hazardous waste collection, pesticide controls, and street sweeping.

- Public education, such as inlet stenciling, public school lessons and programs, public service advertisements, neighborhood meetings, and roadside signs.

- New ordinances that prohibit lawn-clipping disposal, require pet litter pickup, or regulate drainage easements.

- Requirements that developers comply with storm water regulations and incorporate erosion and sediment controls into their plans.

- Requirements that control storm water runoff from municipal properties and operations.

- Elimination of illicit connections and illegal discharges into the storm sewer systems.

The NPDES Program allows local governments to tailor and prioritize their storm water management efforts and local concerns. For example, in Baltimore, Maryland, excessive levels of nutrients are among the city's major water quality concerns as a result of the Chesapeake Bay Program. In spite of significant efforts of BMP implementation over more than twenty years, the city has not met its goal of reducing nutrients discharged to the Bay by 40 percent by the year 2000. The overall Chesapeake Bay Program has not met this goal, but hopes to over the next several years.

The Boston Water and Sewer Commission is concerned about runoff from roadways, especially runoff containing salt and sand in the winter and copper and zinc from automobiles. In September 2000, the Commission began a three-year program to develop and implement a citywide catch basin inspection, cleaning, and preventative maintenance program. The program also includes the development of a database and map linked to the existing GIS.

In responding to a TMDL for trash in the Los Angeles River Watershed, Los Angeles County will eliminate trash in runoff over a ten-year period. The county is testing a variety of devices that remove trash from runoff and specialized catch basins that prevent trash from entering the sewer.

Milwaukee, Wisconsin changed its monitoring and public education activities to test the effectiveness of a BMP targeting public education efforts to a specific community. The BMP includes a monitoring program aimed at the community, the watershed, and the city employees that work in the area.

Worcester, Massachusetts has a significant problem with illicit connections and with flow into the sewers during dry weather. Of its 71 storm water outfalls, 32 of them had drainage areas that carried both sanitary sewage and storm drainage in separate conduits through common manholes. The City retrofitted two thirds of the manholes to prevent the sewage from mixing with the storm water.

Local government costs in complying with the Phase I program generally have bee reported as high. Because of significant inconsistencies in cost accounting and reporting practices, the GAO could not determine the costs to any of the five visited municipalities. Even though all permitted municipalities are required to report cost information annually, EPA has not issued cost reporting guidelines. Consequently, the visited municipalities each reported their program costs differently. A few cities did not report costs for some program elements that served other purposes. Some cities reported costs incurred only by the department that held the permit. Other cities reported only operational costs and neglected all construction costs. Still others did not report costs that were funded by state grant monies. Overall the reported storm water control program costs probably are too low. Some reported program costs are significantly too low.

Officials in all five visited municipalities thought that their annual costs would only increase. A number of factors were mentioned, including

- Additional requirements from watershed management master plans under development.
- Additional requirements imposed by state regulatory agency in future permits.
- Future requirements from TMDLs or other numeric effluent limitations.
- Structural and other deferred BMPs.

None of the five cities had obtained federal funds for their storm water management efforts. Rather, they relied upon local sources, including general revenues, bonds, revenues from specifically created storm water utilities, state grants, and inspection and permit fees.

Conclusions

EPA, state, and municipal officials generally agree that the NPDES Storm Water Program will improve water quality. The officials believe that the program will result in more water bodies that meet water quality standards, improve aesthetic conditions, reduce risk from bacterial contamination, and result in other improvements attributable to the discovery and management of pollutants that otherwise would have gone unnoticed.

However, little hard information is available on the benefits of the storm water program or its general effectiveness. There is little doubt that the results from the Phase I program will become manifest only over time. EPA notes that water quality management programs produce long-term results. The agency expects water quality improvements resulting from Phase I efforts to become evident in the future as the program matures. EPA has not established measurable goals for the overall program. In addition, the agency has not attempted to evaluate the effectiveness of the program in reducing storm water pollution or to determine its cost. The agency attributes this to inconsistent data reporting from permitted municipalities, insufficient staff resources, and other competing priorities within the Office of Wastewater Management.

Although Phase I municipalities report monitoring and cost data annually, the information has not been reviewed to determine whether it can be used to measure the overall program effectiveness or cost. The GAO analysis indicates that the cost data would be difficult or impossible to analyze unless EPA set guidelines for more standardized reporting. Better data on cost and program effectiveness are needed, especially in light of the Phase II program that involves thousands more municipalities in 2003.

Recommendations

The GAO recommends that the EPA

- Establish measurable goals for the program.
- Establish guidelines for obtaining consistent and reliable data on cost and effectiveness.
- Review data submitted by existing permittees to determine whether program goals are met and to identify program costs.
- Assess whether the agency has allocated sufficient resources to oversee and monitor the program nationally.

Sanitary-Sewer Overflow Control Strategy

Richard Field[1], M. ASCE

Abstract

This paper presents a strategy for the abatement of pollution from storm-generated sanitary-sewer overflows (SSO). Because of the great lengths of sanitary-sewer systems and their associated vast number of house-service laterals or building connections, it is often less expensive and more feasible to use alternatives to sewerline rehabilitation for infiltration/inflow (I/I) and SSO control. Extensive sanitary-sewer-system rehabilitation can: (1) be relatively costly; (2) take a long time to complete; and (3) be extremely disruptive to traffic, businesses, and property owners. I/I control studies have found that just correcting I/I in street sewers will not necessarily correct the problem because building connections contribute as much as 60% of the infiltration load. Building connection rehabilitation can be infeasible economically because of the shear number of connections or impractical because of the difficulty in dealing with property owners. Inflow elimination or reduction, cost-effective sewer rehabilitation, and collection system inspection with associated clean out and repair must be performed in all cases. In addition, an integrated economic and feasibility analysis using a combination of maximizing flow to the wastewater treatment plant and maximizing treatment capacity must be considered for controlling the remaining SSO. SSO can be abated using combined-sewer overflow (CSO) control methods. SSO and CSO are both mixtures of municipal sewage, stormwater, and ground water.

Keywords: Sanitary-sewer overflow; SSO; infiltration; inflow; I/I; sewer rehabilitation; house laterals; capacity, management, operation and maintenance (CMOM).

Introduction

This paper presents a strategy for the abatement of pollution from storm-generated sanitary-sewer overflows (SSO) based on research and demonstration projects conducted since the 1970's under the U.S. Environmental Protection Agency's (US EPA's) Urban Watershed Management Branch (UWMB) (formerlythe Storm and Combined Sewer Overflow Research Program). The basic strategy and concepts presented are the authors' recommendations for an economically feasible organized approach to SSO control and as such offers advancements to the field. The

1.Leader, Wet-Weather Flow Research Program, U. S. Environmental Protection Agency, National Risk Management Research Laboratory, Water Supply and Water Resources Division, Urban Watershed Management Branch, Edison, NJ 08837, USA

paper was written in an attempt to assist the US EPA Office of Water (OW) in its current campaign to institute and implement a national SSO rulemaking effort, but does not represent US EPA policy. The campaign started in 1995 with the National Conference on Sanitary Sewer Overflows (US EPA 1996), and the establishment of the SSO Subcommittee to the Urban Wet Weather Federal Advisory Committee and continues with the implementation of the Capacity, Management, Operation and Maintenance (CMOM) Programs for Municipal Separate Storm Sewer Systems (MS4), which was circulated in draft form, in 1999 and 2000, respectively.

Background

SSO and wastewater treatment plant (WWTP) overloads are usually the result of exceeding interceptor and/or WWTP capacity due to infiltration/inflow (I/I); increased population and water consumption in the service area above design allotments; clogged, blocked, and/or silted interceptors; and pumping station failures and other WWTP malfunctions. Disconnecting roof drains and sump pumps will result in significant inflow elimination, while other issues such as high groundwater tables or seasonal impacts from snowmelt may require other approaches. Historically, the UWMB's state-of-the-art (American Public Works Association [APWA] 1970) and manual of practice (APWA 1971) documents identified a significant problem which lead to national emphasis on I/I control and countermeasure research. The UWMB also produced other important I/I documents:

- Infiltration-Inflow Analysis (Cesareo and Field 1975)
- Sewer Infiltration and Inflow Control Product and Equipment Guide (Foster and Sullivan 1977)
- Economic Analysis, Root Control, and Backwater Flow Control as Related to Infiltration/Inflow Control (Sullivan *et al.* 1977a)
- Sewer System Evaluation, Rehabilitation and New Construction — A Manual of Practice (Sullivan *et al.* 1977b)
- Assessment of Sewer Sealants (Sullivan et al. 1982)
- Selected Topics Related to Infiltration and Inflow in Sewer Systems (Sullivan and Ewing 1985)

Other technological developments supported by the UWMB have included detection methodology (Cronk 1972, Pisano *et al.* 1983, Steketee 1985) and instrumentation, preventive installation approaches (Mayer *et al.* 1972), structural materials (The Western Company 1971, Ludwig 1980, Giordano 1978) and construction techniques (American Society of Civil Engineers 1969, Southwest Research Institute 1971, Beetschen and Henry 1978, Olson 1985, Carcich *et al.* 1972) and corrective techniques (The Western Company 1969a, Hoffman and Fiske 1969, Cronk 1972, Driver and Olson 1983, Steketee 1985).

An SSO control program should always consider inflow elimination or reduction, cost-effective sewer rehabilitation, and collection system inspection with associated clean out and repair. In addition, an integrated economic and feasibility analysis using a combination of maximizing flow to the WWTP and maximizing treatment capacity must be considered for controlling any remaining SSO. The collection system components should be properly identified as sanitary, combined, or stormwater. It does not make sense to identify components as the more stringently regulated sanitary if the system is really comprised of combined sewer components that should be permitted as combined-sewer overflow (CSO) and not SSO.

Often, because of the great lengths of sanitary-sewer systems and the shear number of associated building connections, it can be less expensive to use alternatives to sewerline rehabilitation for I/I and associated SSO control. Extensive sanitary-sewer-system rehabilitation alone can:

- be relatively costly;
- take a long time to complete;
- be extremely disruptive to traffic, businesses, and property owners; and
- be ineffective.

I/I control studies have found that just correcting I/I in street sewers will not necessarily correct the problem; some cities estimate that as much as 60% of overflows comes from the house-service laterals or building connections (http://www.epa.gov/owm/sso.htm). Building connection rehabilitation can be economically infeasible because of the shear number of connections or impracticable because of the difficulty in dealing with property owners.

SSO can be controlled using a combination of storage and treatment techniques, just as is the case with combined sewer overflows (CSO). SSO and CSO are both mixtures of municipal sewage and stormwater although the average sanitary sewage to stormwater fraction is higher for SSO. SSO can be abated using CSO control methods.

Design Flow and Monitoring

Design flowrates and volumes should be determined by either long-term simulation or statistical modeling. SSO quantity and quality monitoring should begin immediately and should be used in conjunction with models for either their development, improvement, or calibration and verification. Flow estimates for future I/I reduction or addition should be subtracted or added to design flow values. An advanced collection system modeling approach for capacity assurance of various components of a collection system will allow the best placement of control and treatment alternatives.

Samples should be analyzed for particle-settling velocity and size distribution and dissolved solids fraction as related to pollutants of concern. This will provide the necessary data for more appropriate treatment process selection and design. In addition, correlating historical WWTP quality data with rainfall records could assist in the development of a wet-weather flow (WWF) quality database.

A state-of-the-art systemwide collection system modeling approach has been employed by the County of Henrico, Virginia for planning its SSO control facilities (Lai *et al.* 2000). The Facilities Plan contains a capital improvement program (CIP) for the County's wastewater system through the year 2040. The CIP is divided into phases to accommodate the County's budget and provide flexibility in the future for constructing the improvements when they are needed.

Simply stated, there are two major steps for preparing a facilities plan to mitigate SSO under the present and projected future conditions: (1) identify sewer surcharge segments and potential SSO points in a sewer system and (2) develop and evaluate alternatives to eliminate/control sewer surcharges and SSO. Henrico County accomplished these goals by constructing and applying the EPA SWMM EXTRAN block to the County's sewer system to analyze the capacity of the existing collection system, including pump stations, and identify new facilities needed to meet the anticipated growth and development throughout the County (Lai *et al.* 2000).

Moffa *et al.* (2000) described a retrofit to a sanitary-sewer system to eliminate SSO from an existing regulator structure. SSO from the regulator were caused by rainfall due to a well-established I/I problem within the collection system. A rehabilitation program eliminated the primary I/I sources and regulator SSO.

Sewer-system modeling was performed using the EPA's SWMM Version 4. The previously calibrated model was expanded to identify flow conditions at the regulator following the implementation of the I/I rehabilitation program. The model identified peak overflow rates, volumes, and hydraulic grade lines for various overflow abatement alternatives.

A two-yr frequency, two-h duration design storm was selected in order to determine what control was necessary in terms of treatment, transportation, and/or storage of existing regulator overflows. A simplified model was developed to simulate hourly rainfall for the hypothetical collection system. The data used for these projections consisted of 15 years of hourly rainfall data for the years 1961 through 1985. These data were used as input to the simplified EPA SWMM RUNOFF block that projected flows that were used to simulate long-term overflows using the SWMM TRANSPORT block. The TRANSPORT block was constructed as a simplified network consisting of hypothetical overflows and regulator pipes. The simplified model was used to project the annual volume presently discharged and the percent of annual volume that could be captured by the proposed storage facility. The storage volume, based on SWMM projections, was sized for a two-yr, two-h storm peak overflow rate of 28,000 m^3/d (7.4 MGD) that would discharge a volume of 6,400 m^3 (1.7 MG). This would provide 75% capture of the annual WWF volume compared to that presently discharged without treatment. Retrofitting to capture SSO with a storage facility can provide a cost-effective alternative in lieu of expansion of the sanitary-collection system to convey the excess I/I volume to the treatment plant for primary and secondary treatment. This alternative requires the availability of a sufficient amount of land to build such a facility.

Maximizing Flow to the WWTP

Maximizing flow to the WWTP can be accomplished by two basic methods: (1) utilizing storage facilities and/or (2) increasing interceptor flow-carrying capacity. These methods are based on previously demonstrated and evaluated CSO and/or SSO projects.

Storage facilities. Because of the high volume and variability associated with I/I, additional storage must be considered as a control alternative for SSO. Storage is the best documented WWF abatement measure. In a similar manner as is applied for CSO, SSO is captured and released to the WWTP during low-flow, dry-weather periods. Project results and theory indicate that additional storage capacity must be considered at all times in system planning as it allows for maximum use of existing dry-weather and other treatment plant facilities and results in the lowest cost in terms of pollutant removal. Additional storage facilities possess the favorable attributes of being simple in structural design and operation responding without difficulty to random storm behavior. Frequently they can operate in concert with the regional WWTP. Disadvantages are their large size and land requirements. An important part of SSO control system planning is the determination of the economic break-even point between the required storage volume versus treatment capacity.

The multi-purpose functions of SSO storage facilities should be used wherever possible and include: sedimentation treatment for excessive SSO volumes greater than storage volume, dry-weather flow (DWF) equalization , flood protection, and sewerline relief. Sedimentation treatment in storage can either relieve a WWTP solids burden or remove a significant pollution fraction of an excessive WWF which causes a storage facility overflow.

Storage can be applied upstream, midstream, or downstream at the SSO point (either at the remote SSO location or at the WWTP). Upstream storage can offer the dual benefit of drainage and flood relief/control. Types of storage facilities include: unused capacity in the existing interceptor and sewerlines, conventional concrete tanks and lined earthen basins, the minimum land requirement tunnels and underground tanks, natural and mined under- and above-ground formations, and abandoned facilities. Compartmentalizing of the storage basin for DWF capture may reduce sludge handling problems. New sewers can be made larger for WWF conveyance and storage, e.g., Great Britain typically designs for six times DWF while the United States uses twice DWF. In-sewer storage and use of abandoned facilities for storage will be the least expensive types of storage and therefore should be considered first. Earthen basins should be considered next since they are relatively inexpensive.

Minneapolis (Metropolitan Sewer Board 1971), Detroit (Watt *et al.* 1975), and Seattle (Leiser 1974) made use of unused storage capacity within the existing sewerage system for the purpose of reducing the frequency and volume of CSO. The City of Rohnert Park, California (Welborn 1974), as part of a demonstration of the dual use of storage tanks for equalizing peak DWF as well as WWF, illustrated the economic advantages of their use as part of a regional wastewater handling scheme. However, an evaluation (Moffa *et al.* 2000) of a small sanitary-sewer system, showed marginal cost benefit with reduced cumulative annual overflow.

Interceptor flow-carrying capacity. Interceptor flow carrying capacity can be increased by:
1. cleaning out clogged and silted-up interceptors,
2. increasing the pumping capacity for surcharged interceptors,
3. installing larger or parallel interceptors, or
4. applying polymer injection or lining to reduce pipe roughness.

The inspection of interceptors for blockages and siltation and associated clean out should be done first, since this is apt to be the least expensive way of gaining flow capacity and is part of the CMOM. Items 2 and 3 are typical steps in areas experiencing rapid development.

UWMB research has shown that item 4, polymer injection, can increase flow capacity as much as 2.4 times at a constant head (The Western Company 1969b). This method can be used as a short-term measure to correct troublesome localized flooding and SSO prior to or during system expansion. In Dallas, Texas (Chandler and Lewis 1977), the existing sewerlines were relatively inaccessible but capable of handling normal DWF. An ongoing I/I abatement program was already initiated, however, a quick low-cost solution was needed to eliminate SSO in the interim. Portable and manned polymer injection systems and the use of either dry or slurry polymer were demonstrated to reduce or prevent SSO.

Maximizing WWTP Treatment Capacity

Approaches to accomplish maximizing WWTP capacity include:

- increasing hydraulic loading to the existing treatment process,
- retrofitting primary settling tanks, or
- installing parallel processes.

Increasing hydraulic loading. Increasing hydraulic loading (HL) to the existing primary/secondary treatment processes without modifications should be based on stress testing (to determine removal efficiency versus HL). This is the least expensive method and should be considered first.

Retrofitting primary settling tanks. Retrofitting primary settling tanks can be accomplished with plate settlers, chemical addition facilities, or dissolved air floatation (DAF). Consideration of an automated switching process should be made to converting from settling during low-flow to DAF during high-flow conditions to save power costs (Benedict and Roelfs 1981).

Installing parallel processes. Parallel process train scenarios include:

- primary physical/chemical treatment at the WWTP,
- secondary biological treatment for new WWTP installations, and
- disinfection.

Depending on the level of treatment, impacts to the receiving water, and agreements with the local permitting authority, the parallel process train effluent can be either discharged directly to the receiving water, directed back to the WWTP influent, or both, by means of flow splitting. If it is directed to the influent, then a flow equalization basin could be needed and disinfection would not be needed.

Primary Physical/Chemical Treatment at the WWTP. Options for primary physical/chemical parallel processes include: bar screening followed by aerated or swirl degritters followed by microscreens, plate settlers, fine-mesh screening/dual-media high-rate filters, fine-mesh screening/DAF, chemically-enhanced primary treatment (CEPT), microcarrier-enhanced settling, or conventional primary settlers. The benefit of a swirl degritter over a conventional aerated degritter is that they have no moving parts (and therefore no external power requirement) and require less space.

The Bachman WWTP in Dallas, Texas (Wolf 1977) suffered from excessive I/I. Accordingly, tube settlers were retrofitted into the existing primary settling tanks in order to afford significantly greater WWF HL. Although this settling concept works on a theoretical basis there was insufficient WWF to make an evaluation. Plate settlers, like tube settlers, have been shown to allow five or more times the HL of conventional settling tanks. Plate settlers are angled plates placed one on top of the other to greatly increase the surface settling area. Both tube and plate settlers have widespread industrial application.

Microsceening (Moffa 1989 and Lager *et al.* 1977) has been demonstrated to be effective for CSO primary treatment in Philadelphia, Pennsylvania (Maher 1974) and Syracuse, New York (Drehwing *et al.* 1979). In Philadelphia a 23 Φm aperture microscreen attained average SS removals of 70% with increased SS removals of 78% using polyelectrolyte addition. The high HL was 39 m/h (16 gpm/ft^2) without and 88 m/h (36 gpm/ft^2) with polyelectrolyte addition. The Department of Drainage and Sanitation, Onondaga County, New York demonstrated a Zurn™ 71 Φm and a Crane™ 23 Φm aperture microscreening unit, respectively. Average SS removals were 48% at HLs between 12 and 152 m/h (5 and 62 gpm/ft^2) and 58% at HLs between 8.1 and 33 m^3/m^2/h (3.3 and 13.7 gpm/ft^2), respectively.

A 30 in. diameter pilot-scale high-rate filter (HRF) was sucessfully demonstrated for the dual-purpose treatment of sanitary and combined sewage (Innerfeld *et al.* 1979). As previously indicated, CSO and SSO have the same components; they are basically mixtures of sanitary sewage and stormwater. The system included a pretreatment 420 Φm screen to enable longer filter run times. A deep-bed dual-media filter of either 1.2 or 1.5 m (48 or 60 in.) of anthracite over 0.76 m (30 in.) of sand was used. The deep bed with the coarser lower specific gravity anthracite on top allowed greater pollutant capture effectiveness, longer run times, and effective backwashing. Chemical addition was also demonstrated. The HRF only requires 5 to 7% of the area of conventional sedimentation.

Table 1 shows weighted average removals for CSO and sanitary sewage with filter HLs of 39 m^3/m^2/h (16 gpm/ft^2) and 20 and 29 m^3/m^2/h (8 and 12 gpm/ft^2), respectively. The table does not distinguish between the separate CSO test runs which employed no chemical addition and chemical addition.

Table 1. Dual DWF/WWF Treatment, Newtown Creek WWTP, NYC

Treatment	Hydraulic Loading	Percent System Removals (%)		
	m^3/m^2/h	Suspended Solids	Biochemical Oxygen Demand (5-d)	Chemical Oxygen Demand
Combined Sewage	39	66	41	47
Sanitary Sewage	20, 29	67	42	44

CEPT uses metal salts (electrolytes) and polymers (polyelectrolytes) to enhance settling. Morrissey and Harleman (1992) cite several successful applications of polyelectrolytes reducing the use of metal salts. Sludge volume for the Hyperion Treatment Plant of San Diego, California only increased 10% and 6%, respectively, before and after digestion. CEPT at Hyperion attained removals of 83% SS and 52% BOD_5 at an average HL of 3 $m^3/m^2/h$ (1.2 gpm/ft^2) with additions of 20 mg/L of ferric chloride and 0.25 mg/L of anionic polymer.

Microcarrier-enhanced settling or ballasted-flocculating settling systems can also be installed as part of a treatment train. Microcarrier systems such as the Actiflo™ (Vivendi), use recycled microcarrier (e.g., microsand [0.05–0.2 mm]), while DensaDeg™ (Infilco Degremont) uses recycled sludge as the carrier to enhance settling. Each system adds chemicals to the recycled material to create heavier/larger floc which does not require as long a detention time for settling as conventional settling tanks. The Actiflo™ claims 80% removals of SS with effluents of 30 mg/l, a 15 min detention time, an ability to handle variable HLs, and a peak HL of 130 $m^3/m^2/h$ (52 gpm/ft^2) as compared to 1 $m^3/m^2/h$ (0.5 gpm/ft^2) for conventional settling tanks. The DensaDeg™ system claims HLs from 17 to 24 $m^3/m^2/h$ (10 to 14 gal/ft^2) and a 50% reduction in unit size as compared to conventional solids contact clarifiers with the added benefit of incorporating gravity-sludge thickening. Performance of microcarrier-enhanced settling may require pretreatment, i.e., screening in excess of bar screening and surge tanks, especially for offline situations, or rapid startup.

Secondary Biological Treatment for New WWTP Installations. Secondary biological treatment options for new WWTP installations include: high-rate trickling filtration (deep honeycomb plastic media) having two filters in parallel during WWF operation and automatically switching back to series during DWF operation (Homack *et al.* 1973); contact stabilization during WWF periods automatically switching to conventional activated sludge during DWF periods (Benedict and Roelfs 1981 and Agnew *et al.* 1975); rotating biological contactors (shaft-mounted rotating disks) (Welsh and Stucky 1974); and biological fluidized-bed filtration (Scheible and Grey 1988). These processes have been demonstrated to treat SSO or WWF at relatively high hydraulic and organic loadings.

Retrofitting was conducted at the trickling filter WWTP in New Providence, New Jersey to alleviate hydraulic overloading and resultant loss of treatment efficiency caused by excessive sanitary sewer infiltration (Homack *et al.* 1973). The plant utilized two trickling filters. The older existing filter was comprised of conventional rock media, and the newer, a high-rate retrofit, comprised of deep-bed honeycomb-plastic media, operated in parallel to treat WWF. During DWF periods the plant was operated in series to maintain an active biological slime on the filters. Series operation also served to improve DWF treatment efficiency. The trickling filter operation automatically switches from series to parallel when the WWF exceeds 0.12 m^3/s (2.8 MGD) and reverses back to series when the flow goes below 0.09 m^3/s (2.0 MGD).

The New Providence investigation has shown that it is both technically feasible and economical to construct and operate a treatment plant to process both DWF and the higher flows encountered during periods of excessive infiltration using a series-parallel combination of high-rate trickling filters. Table 2 indicates that DWF BOD_5 and SS removal efficiencies were approximately 93%, respectively. The corresponding concentrations of BOD_5 and SS in the final effluent were 9 mg/L and 12 mg/L, respectively. WWF BOD_5 and SS removal efficiencies were approximately 86%, respectively.

Table 2. Dual DWF/WWF WWTP, New Providence, New Jersey

Type of Flow	Biochemical Oxygen Demand (5-d)		Suspended Solids	
	Effluent concentration (mg/L)	Percent Removal (%)	Effluent concentration (mg/L)	Percent Removal (%)
Dry-weather Flow	9	94	12	93
Wet-weather flow		87		86

A drawback of the high-rate trickling filter process for the treatment of high flows is the tendency for colloidal suspended material to be discharged from the filters. Unless chemical precipitation and low secondary clarifier overflow rates or supplemental biological flocculation in the form of a pond are provided, high SS and BOD_5 removal will not be obtained. It was found that the HL should be limited to 2.73 $m^3/m^2/h$ [70 million gallons per acre per day (MGAD) and organic loading rate to 1.4 kg/m^3 (85 lb BOD_5/1000 ft^3) on the plastic media filter and 0.78 $m^3/m^2/h$ (20 MGAD)] and 0.64 kg/m^3 (40 lb BOD_5/1000 ft^3) on the rock media filter to achieve an effluent containing 25 mg/L BOD_5 or less during periods of excessive infiltration.

The plastic media trickling filter removed about 2.7 times more BOD_5 than the rock media. Accordingly, it was recommended that the plant operate at a 2:1 flow split between the plastic: rock media filters during WWF. While the initial cost of the plastic media filter was 25% more than the rock, the overall ratio of cost per kg of BOD_5 removed per m^3 was 2:1 in favor of the plastic media type.

In Clatskanie, Oregon, a full-scale dual-use facility was constructed to alleviate SSO caused by excessive infiltration (Benedict and Roelfs 1981). The plant treated WWF as high as 6.6 times the DWF of 0.008 m^3/s (0.19 MGD). The plant's primary physical and secondary biological treatment processes were both modified to convert automatically from conventional sedimentation and activated sludge treatment during DWF periods to DAF and contact stabilization during WWF periods since DAF and contact stabilization can accommodate higher Hls. DWF BOD_5 and SS removal efficiencies were 94%, respectively and WWF BOD_5 and SS removal efficiencies for both short- and long-term storms were 73% and 71%, respectively (Benedict and Roelfs 1981).

Contact stabilization sludge treatment relies on the sorptive ability of biological solids to remove biodegradable organic matter in short retention periods (Benedict and Roelfs 1981). These solids are settled out in the secondary clarifier, returned to the aerobic digester for stabilization and recycled to mix with the influent of the contact chamber. The recycled flow determines the contact chamber time and aerobic stabilization time. The contact, aeration, and sludge reaeration compartments are further subdivided into operating zones to provide flexibility. In general, aeration volume requirements of contact stabilization (a completely mixed system) are typically 50% less than conventional plug flow (Tchobanoglous and Burton 1991). The DAF-contact stabilization capital and total costs were only 14% and 10%, respectively higher than the costs of a standard DWF plant. In Clatskanie, it was estimated that dual treatment DWF and WWF as an I/I control method costs less than half as much as either sewer rehabilitation or flow equalization.

A CSO demonstration project in Milwaukee, Wisconsin evaluated rotating biological contactors consisting of a series of shaft-mounted rotating disks (Welsh and Stucky 1974). Removals can exceed that of trickling filters; however a flow surge facility appeared essential.

Disinfection. If the existing unit cannot provide adequate disinfection, parallel or retrofitted high-rate disinfection processes are recommended because they require smaller contact tank volume. High-rate disinfection processes that have been successfully demonstrated to work for CSO include (Lager *et al.* 1977, Moffa, Ed. 1989, and O'Shea and Field 1992):

- increased mixing intensity;
- increased disinfectant concentration;
- more rapid disinfection oxidants;
- two-stage dosing; or
- combinations of the above.

Ozone and chlorine dioxide are relatively rapid oxidants. Ozone also supplies additional oxygen to the wastewater. High-intensity mixing is accomplished by mechanical-flash mixers at the disinfection point. Slower static mixing using vanes and narrow corrugated pathways should also be considered. Adequate SS concentration and particle size reductions are necessary for satisfactory microorganism kill (O'Shea and Field 1992).

High chlorine dosing, while effective, creates toxic and carcinogenic residuals. Sequential addition of chlorine followed by chlorine dioxide at intervals of 15 to 30 s enhanced high-rate disinfection beyond the expected additive effect. A minimum effective combination of 8 mg/L of chlorine followed by 2 mg/L of chlorine dioxide was as effective as 25 mg/L of chlorine alone.

Advanced Technologies

Innovative, advanced, and evolving treatment technologies should also be considered. For example, two high-rate treatment systems have been demonstrated to provide high levels of pollution removal efficiency for CSO, and therefore should be applicable for the treatment of SSO.

In Albany, New York (Shuckrow *et al.* 1973 and Moffa, Ed. 1989) a 0.1 MGD pilot powdered activated carbon-alum coagulation unit was evaluated for treating sanitary and combined sewage. Both powdered carbon and alum were added in a static mixing/reaction pipeline and the resultant coagulated matter was flocculated downstream and separated by tube settlers. Average removals in excess of 94% COD and BOD_5 and 99% SS were consistently achieved in treating CSO.

High-gradient magnetic separation (HGMS) is an advanced treatment technology that has been applied using a pilot system for combined and sanitary sewage in Boston, Massachusetts (Allen *et al.*1977). HGMS consists of an iron-framed canister packed with a fibrous ferromagnetic material that is magnetized by coils surrounding the canister. HGMS may be used to remove nonmagnetic contaminants by binding magnetic seed particles, e.g., magnetite with or without coagulants to the contaminants. HGMS can provide rapid filtration at 245 $m^3/m^2/h$ (100 gpm/ft^2). The Boston pilot demonstrated a high level of pollution removal including: 95% SS, 92% BOD_5, 74% COD, 99.2% fecal coliform, and 99+ % virus, respectively.

Satellite Treatment

Satellite treatment should be comprised of automatic physical processes with remote monitoring to reduce manpower requirements. Maintenance issues will be more pronounced and potential failures at remote locations along with associated water quality impacts will exist. Location issues such as sensitive receiving stream segments and neighborhood aesthetics should be considered. Accordingly, physical treatment processes at satellite locations should only be used if the WWTP cannot handle all of the WWF. Satellite storage should account for dual-purpose treatment by settling for WWF volumes greater than the storage volume.

Recommended National Requirements

Inflow elimination or reduction, cost-effective sewer rehabilitation, and collection system inspection with associated clean out and repair must be performed in all cases; and an integrated economic and feasibility analysis using a combination of maximizing flow to the WWTP and maximizing WWTP capacity must be considered for controlling the remaining SSO.

In the authors' opinion, National SSO treatment requirements should be a technology-based approach coupled with more stringent human health and receiving-water quality risk-based requirements. In other words, if the applied treatment technology results in receiving-water quality impairments, then higher treatment levels would be required. Exceptions should be allowed based on the community's showing of (1) no additional water-quality impairment or health risks and (2) economic restraints which would give the community a longer time for compliance but not affect the standard. The US EPA will need guidelines for economic distress and should encourage compliance by a granting mechanism. The technology-based approach should be based on national guidelines with the above-stated exceptions.

Since SSO event volumes and flowrates and long-term volumes and average flowrates are significantly lower than the comparable values for CSO it is felt that it should be economically feasible to eliminate most SSO and in the worst case allow no more than two SSO per year. Consideration must be given to bypassing secondary treatment and just providing partial "primary" treatment by directing the excessive SSO through a storage facility and/or physical treatment and disinfection based on receiving-water quality impacts and BOD_5 dilution in the influent. This strategy is one of the nine minimum CSO control measures presented in, "Maximization of Flow to the POTW for Treatment" (US EPA 1995).

Conclusions

Control technology must be considered for SSO control in addition to inflow reduction, cost-effective sewer rehabilitation, and sewer system inspection and associated clean out and repair. Conducting an economic/feasibility analysis of all alternatives that maximize flow to the WWTP and its treatment capacity will determine whether I/I control or alternative methods are more cost effective and/or more practicable. SSO control methods include:

- removing inflow;
- performing only cost-effective sewer rehabilitation on a continuing basis;
- modeling system response for planning and design using the EPA SWMM EXTRAN block which predicts surcharging;
- inspecting the interceptor, pumping station and sewer system for blockages, siltation and structural failures which cause SSO (and cleaning out or repairing as necessary);
- maximizing flow to the WWTP by using storage facilities and/or increasing interceptor flow-carrying capacity;
- maximizing the treatment capacity of the WWTP capacity by retrofitting, installing parallel processes, or constructing new high-rate WWTP; and
- installing satellite treatment facilities.

References

Agnew, R. W. *et al.* (1975), Biological Treatment of Combined Sewer Overflow at Kenosha, Wisconsin, US EPA, ORD, Cincinnati, OH, EPA-670/2-75-019, NTIS PB 242 126.

Allen, D. M., *et al.* (1977), Treatment of Combined Sewer Overflows by High gradient Magnetic Separation, US EPA,ORD, Cincinnati,OH, EPA-600/2-77-015, NTIS PB264 935.

American Society of Civil Engineers (1969), "Combined Sewer Separation Using Pressure Sewers, US EPA, ORD, Washington, DC, 11020EKO10/69, NTIS PB 188 511.

American Public Works Association (1970), Control of Infiltration and Inflow into Sewer Systems, Federal Water Pollution Control Administration, 11022EFF 12/70, NTIS PB 200 827.

American Public Works Association (1971), Prevention and Correction of Excessive Infiltration and Inflow into Sewer Systems—A Manual of Practice , US EPA, ORD, Washington, DC, 11022EFF01/71, NTIS PB 203 208.

Beetschen, L.J. and W.C. Henry (1978), Evaluation of Trenchless Sewer Construction at South Bethany Beach, Delaware, US EPA, ORD, Cincinnati, OH, EPA-600/2-78-022, NTIS PB 278 776.

Benedict, A. H. and V. L. Roelfs (1981), Joint Dry/Wet Weather Treatment of Municipal Wastewater at Clatskanie Oregon, US EPA, ORD, Cincinnati, OH, EPA-600/2-81-061, NTIS PB 81-187 262.

Carcich, I.G. *et al.* (1972), A Pressure Sewer System Demonstration , US EPA, ORD, Washington, DC, EPA-R2-72-091, NTIS PB 214 409.

Cesareo, D. J., and Field, R. (1975), Infiltration-Inflow Analysis, Journal of the Environmental Engineering Division, ASCE, Vol. 101, No. EE5, Proc. Paper 11645. October, 1975, pp. 775-785.

Chandler, R. W. and Lewis, W. R. (1977), Control of Sewer Overflows by Polymer Injection, US EPA, ORD, Cincinnati, OH, EPA-600/2-77-189, NTIS PB 272 654.

Cronk, G.E. (1972), Ground Water Infiltration and Internal Sealings of Sanitary Sewers, Montgomery County, OH, US EPA, ORD, Washington, DC, 11020DHQ 06/72, NTIS PB 212 267.

Drehwing, F. *et al.* (1979) Disinfection/Treatment of Combined Sewer Overflows, Syracuse, New York, US EPA, ORD, Cincinnati, OH, EPA-600/2-79-134, NTIS PB 80-113 459.

Driver, F.T. and M.R. Olson (1983), Demonstration of Sewer Relining by the Insituform Process, Northbrook, Illinois, US EPA, ORD, Cincinnati, OH, EPA-600/2-83/064, NTIS PB 83-245 878.

Foster, N. S., and Sullivan, R. H. (1977), Sewer Infiltration and Inflow Control Product And Equipment Guide, US EPA, ORD, Cincinnati, OH, EPA 600/2-77-017c, NTIS PB 272 680.

Giordano, J.J. (1978), The Construction, Technical Evaluation, and Frictional Determination of an Aluminum Storm Sewer System, US EPA, ORD, Cincinnati, OH, EPA 600/2-78-025, NTIS PB 283 458.

Hoffman and Fiske, (1969), Reduction of Groundwater Infiltration into Sewers by Zone Pumping at Meridian, Idaho, Federal Water Pollution Control Administration, 11020---06/69, NTIS PB 187 868.

Homack, P.H. *et al.* (1973), Utilization of Trickling Filters for Dual Treatment of Dry and Wet Weather Flows, US EPA, ORD, Washington , DC, EPA-670/2-73-071, NTIS PB 231 251.

Innerfeld, H. *et al.* (1979), Dual Process High-Rate Filtration of Raw Sanitary Sewage and Combined Sewer Overflows, US EPA, ORD, Cincinnati, OH, EPA-600/2-79-015, NTIS PB 296 626.

Lai, F. *et al.* (2001), Planning for SSO Control: Henrico County, VA - Case Study, Environmental and Water Resources Institute's (EWRI's) World Water & Environmental Resource Congress, May 20-24, 2001, Orlando, Florida.

Lager, J. A. *et al.* (1977), Urban Stormwater Management and Technology--Update and Users' Guide, US EPA, ORD, Cincinnati, OH, EPA-600/8-77-014, NTIS PB 275 654.

Leiser, C.P. (1974), Computer Management of a Combined Sewer System , US EPA ORD, Washington, DC, EPA-670/2-74-022, NTIS PB 235 717.

Ludwig, A.C. (1980), Impregnation of Concrete Pipe for Corrosion Resistance and Strength Improvement, US EPA, ORD, Cincinnati, OH, EPA-600/2-80-009, NTIS PB 80-190 234.

Maher, M.B. (1974), Microstraining and Disinfection of Combined Sewer Overflows -- Phase III, US EPA ORD, Washington, DC, EPA-670/2-74-049, NTIS PB 235 771.

Mayer J.K. *et al.* (1972), Sewer Bedding and Infiltration Gulf Coast Area , US EPA ORD, Washington, DC, 11022DEI05/72, NTIS PB 211 282.

Metropolitan Sewer Board (1971), Dispatching Systems for Control of Combined Sewer Losses, US EPA ORD, Washington, DC, 11020FAQ03/71, NTIS PB 203 678.

Moffa, P., Editor (1989), Control and Treatment of Combined-Sewer Overflows, Chapter 4 - Combined-Sewer Overflows: Control and Treatment, Van Nostrand Reinhold, New York.

Moffa, P. *et al.* (2000), "Retrofitting Control Facilities For Wet-Weather Flow Control," ORD, EPA-600/R-00/020, Cincinnati, OH.

Morrissey, S.P. and Harleman, D.R.F. (1992), Retrofitting Conventional Primary Treatment Plants for Chemically Enhanced Primary Treatment in the U.S.A. Proceedings, 5th Gothenburg Symposium, Nice, France. Chemical Water and Wastewater Treatment II. Springer-Verlag Berlin Heidelberg, pp. 401-416.

Olson, M.R. (1985), Insituform and Other Sewer Rehabilitation Techniques, US EPA, ORD, Cincinnati, OH, EPA-600/2-85/132, NTIS PB 86-130 192.

O'Shea , M.L. and Field, R. (1992), An evaluation of Bacterial Standards And Disinfection Practices Used for the Assessment and Treatment of Stormwater, Advances in Applied Microbiology, Volume 37.

Pisano, W.C. *et al.* (1983), Value of Flow Calibration for Decision Making in Infiltration/Inflow Studies, US EPA, ORD, Cincinnati, OH, EPA-600/2-83/087, NTIS PB 83-259 739.

Scheible, K.C. *et al.* (1986), Ultraviolet Disinfection of Wastewaters from Secondary Effluent and Combined Sewer Overflows, US EPA, ORD, Cincinnati, OH, EPA-600/2-86-005, NTIS PB-86-145 182.

Shuckrow, A.J. *et al.* (1973), Physical-Chemical Treatment of Combined and Municipal Sewage, US EPA, ORD, Washington, DC, EPA-R2-73-149, NTIS PB 219 668.

Southwest Research Institute (1971), Impregnation of Concrete Pipe , US EPA ORD, Washington, DC, 11024EQE06/71, NTIS PB 213 44.

Steketee, C.H. (1985), Demonstration of Service Lateral Testing and Rehabilitation Techniques, US EPA, ORD, Cincinnati, OH, EPA-600/2-85/131, NTIS PB 86-135 647.

Sullivan, R. H., *et al.* (1977a), Economic Analysis, Root Control as Related to Infiltration/Inflow: Volume I - Final Report, US EPA, ORD, Cincinnati, OH, EPA -600/2-77-017a, NTIS PB 280 738.

Sullivan, R. H. *et al.*, (1977b), Sewer System Evaluation Rehabilitation and New Construction - A Manual of Practice, US EPA, ORD, Cincinnati, OH, EPA 600/2-77-017d, NTIS PB 279 248.

Sullivan, R. H. *et al.* (1982), Assessment of Sewer Sealants, US EPA, ORD, Cincinnati, OH, EPA 600/8-82-012, NTIS PB 82 262 718.

Sullivan, R. H. and Ewing, J.W. (1985), Selected Topics Related to Infiltration and Inflow in Sewer Systems, US EPA, ORD, Cincinnati, OH, EPA-600/2-85/020, NTIS PB 85-102 523.

Tchobanoglous, G. and Burton, F.L. (1991), Wastewater Engineering: Treatment Disposal and Reuse, Metcalf and Eddy, Inc. 3rd Edition, McGraw-Hill ,Inc., New York.

US EPA (1995), Combined Sewer Overflows, Guidance for Nine Minimum Controls, Office of Wastewater Management, EPA 832-B-95-003, Washington, D.C.

US EPA (1996), Seminar Publication: National Conference on Sanitary Sewer Overflows (SSOs) US EPA, ORD, Washington, DC. EPA/625/R-96/007.

Watt, T.R. *et al.* (1975), Sewerage System Monitoring and Remote Control, US EPA, ORD, Cincinnati, OH, EPA-670/2-75-020, NTIS PB 242 107.

Welborn , H. L. (1974), Surge Facility for Wet and Dry Weather Flow Control, US EPA, ORD, Cincinnati, OH, EPA-670/2 - 74 - 075, NTIS PB 238 905.

Walesh, S. and Carr, R. (2000) Street Storage System for Control of Combined Sewer Surcharge: Retrofitting Stormwater Storage Into Combined Sewer Systems US EPA, ORD, Cincinnati, OH, EPA/600/R-00/065.

Welsh, F. L. and Stucky, D. J. (1974), Combined Sewer Overflow Treatment by the Rotating Biological Contactor Process, US EPA, ORD, Cincinnati, OH, EPA-670/2-74-050, NTIS PB 231 892.

Western Company, The (1969a), Improved Sealants for Infiltration Control, Federal Water Pollution Control Administration, 11020DIH06/69, NTIS PB 185 950.

Western Company, The (1969b), Polymers for Sewer Flow Control, Federal Water Pollution Control Administration, 11020DIG08/69, NTIS PB 185 951.

Western Company, The (1971), Heat Shrinkable Tubing as Sewer Pipe Joints , US EPA ORD, Washington, DC, 11024FLY06/71, NTIS PB 208 816.

Wolf, H. W. (1977), Bachman Treatment Facility for Excessive Storm Flow in Sanitary Sewers, US EPA, ORD, Cincinnati, OH, EPA-600/2-77-128, NTIS PB 269 128.

Simulation and Analysis of Debris Flow of Chui-Sue River Watershed

Meei-Ling Lin [1], A.M. ASCE, Kuo-Long Wang [2], and Gee-Jeng Huang [2]

Abstract

In 1996 typhoon Herb struck the central Taiwan area, causing severe debris flow in many locations of the Chen-You-Lan river watershed. In order to understand the potential affected area and its severity, it is desirable to be able to simulate the flow route of debris. In this research numerical simulation of debris flow and deposition process had been carried out using FLO-2D adopting Chui-Sue River watershed as the study case. The microgeomorphic database had been created to distinguish the position and volume of debris deposition at Chui-Sue River. Results of the potential analysis were satisfactory on estimating debris flow region and deposition thickness when compared to the field data obtained from microgeomorphic study.

Keywords: debris flow, potential analysis, microgeomorphic study, numerical simulation

Introduction

Due to the vast development of economy, usage and development of the land in mountain terrain has grown rapidly for the available land in plane area is limited. However, the mountain terrain in Taiwan is usually very steep and with fragile geological conditions. Heavy rainfall carried by typhoon often caused severe hazard by inducing slope failure and debris flow. In 1996, typhoon Herb struck Taiwan, dumping almost 2000mm of rainfall. Severe debris flow occurred at the Chui-Sue river watershed several times since then. In order to be able to manage and mitigate the debris flow, it is desirable to be able to simulate the flow route and deposition process of the debris. The purpose of this research is to construct a simulation model for evaluation of potentially affected area and to understand the degree of severity, which could be provided as a base for decision making of hazard mitigation. In this paper, the method for determining the potential affected area

1. Group Leader, Office for National S&T Program for Hazard Mitigation
Prof., Dept. of Civil Engineering, National Taiwan University, Taipei 106, Taiwan
2. Graduate Student, Dept. of Civil Engineering, National Taiwan University, Taipei 106, Taiwan

and impact caused by debris flow was developed based on numerical simulation. The microgeomorphic data of the Chui-Sue river watershed were used in verifying the model obtained from the debris flow caused by typhoon Herb in this research.

Numerical Model of Debris Flow

For the numerical simulation, a commercially available flow model, FLO-2D, was used to analyze the two dimensional two-phase flow. The governing equations involved in developing the model were as presented in the follow.

1. Continuity equation

$$\frac{\partial h}{\partial t}+\frac{\partial[uh]}{\partial x}+\frac{\partial[vh]}{\partial y}=i \qquad (1)$$

where，

h : the flow depth

u, v : depth-averaged velocity components along the x- and y- coordinates

i ： excess rainfall intensity

t ： time

2. 2-D equations of motion

$$S_{fx}=S_{ox}-\frac{\partial h}{\partial x}-\frac{u}{g}\frac{\partial u}{\partial x}-\frac{v}{g}\frac{\partial u}{\partial y}-\frac{1}{g}\frac{\partial u}{\partial t}$$

$$S_{fy}=S_{oy}-\frac{\partial h}{\partial y}-\frac{v}{g}\frac{\partial v}{\partial y}-\frac{u}{g}\frac{\partial v}{\partial x}-\frac{1}{g}\frac{\partial v}{\partial t} \qquad (2)$$

where，

S_{ox}、S_{oy} ： bed slope components

S_{fx}、S_{fy} ： the friction slope components

g ： gravity acceleration

By neglecting the last three terms in the above equations, the approximation of diffuse wave equations could be obtained. While neglecting the pressure gradient term, a kinematic wave equation could be derived. On most steep slopes, usually the application of the kinematic wave would be sufficient to model the floodwave progression.

3. Rheological model(O'Brien and Julien, 1985)

$$\tau=\tau_y+\tau_v+\tau_t+\tau_d$$

$$=\tau_y+\eta\left(\frac{\partial u}{\partial y}\right)+C\left(\frac{\partial u}{\partial y}\right)^2 \qquad (3)$$

where：

τ ： the total shear stress

τ_y ： the yield shear stress

τ_v ： the viscous shear stress

τ_t ： the turbulent shear stress

τ_d ： the dispersive shear stress

η ： the dynamic viscosity

and with C denotes the inertial shear coefficient,

$$C = \rho_m l^2 + f(\rho_m, C_v) d_s^2$$

ρ_m : mass density of the mixture

l : the Prandtl mixing length

d_s : the sediment size

Bagnold(1954)gave the definition of $f(\rho_m, C_v)$ as follow :

$$f(\rho_m, C_v) = a_i \rho_m \left((\frac{C_*}{C_v})^{1/3} - 1 \right) \qquad (4)$$

where ,

a_i : empirical coefficient, which equals to0.01

C_* : the maximum static volume concentration for the sediment particles

in Eq.(3), the first term represents the yield shear stress, which is the Mohr-Coulomb shear resistance, the second term is the viscous shear stress, the third term is the turbulent shear stress, and the fourth term is the dispersive shear stress. The above equation could be depth integrated and rewritten in the slope form :

$$S_f = S_y + S_v + S_{td}$$

$$= \frac{\tau_y}{\gamma_m h} + \frac{K\eta u}{8\gamma_m h^2} + \frac{n^2 u^2}{h^{4/3}} \qquad (5)$$

where ,

S_y : yield slope

S_v : viscous slope

S_{td} : turblent-dispersive slope

γ_m : the specific weight of the sediment mixture

K : resistance parameter

n : equivalent Manning value

The yield stress τ_y and the viscosity η vary principally with sediment concentration, and can be written in empirical relationships as,

$$\tau_y = \alpha_1 e^{\beta_1 C}, \eta = \alpha_2 e^{\beta_2 C} \qquad (6)$$

$$n = n_t + n_d = n_t + 0.0538 n_t e^{6.0896C} \qquad (7)$$

n_t : turbulent Manning value

n_d : dispersive Manning value

The yield and viscosity stresses in Eq. (6) appeared to be function of the volumetric sediment concentration, where the parameters α_1, α_2, β_1, and β_2 are empirical coefficients determined by laboratory experiment. The ranges of values of these parameters were as listed in Table 1.

Table 1 Parameters α_1, α_2, β_1, and β_2 in Yield Stress and Viscosity (O'Brien&Julien,1985)

Source (1)	$\tau_y = \alpha_2 e^{\beta_2 c_v}$ α_2 (dynes/cm^2) (2)	β_2 (3)	$\eta = \alpha_1 e^{\beta_1 c_v}$ α_1 (poises) (4)	β_1 (5)
(a) Relationships Found in Field				
Aspen Pit 1	1.81×10^{-1}	25.7	3.60×10^{-2}	22.1
Aspen Pit 4	2.72	10.4	5.38×10^{-2}	14.5
Aspen nature soil	1.52×10^{-1}	18.7	1.36×10^{-3}	28.4
Aspen mine fill	4.73×10^{-2}	21.1	1.28×10^{-1}	12.0
Aspen natural soil source	3.83×10^{-2}	19.6	4.95×10^{-4}	27.1
Aspen natural fill source	2.91×10^{-1}	14.3	2.01×10^{-4}	33.1
Glenwood 1	3.45×10^{-2}	20.1	2.83×10^{-3}	23.0
Glenwood 2	7.65×10^{-2}	16.9	6.48×10^{-1}	6.2
Glenwood 3	7.07×10^{-4}	29.8	6.32×10^{-3}	19.9
Glenwood 4	1.72×10^{-3}	29.5	6.02×10^{-4}	33.1
(b) Relationships Found in Literature				
Iida (1938)	—	—	3.73×10^{-5}	36.60
Dai et al. (1980)	2.60	17.48	7.50×10^{-3}	14.39
Kang and Zhang (1980)	1.75	7.82	4.05×10^{-2}	8.29
Qian et al. (1980)	1.36×10^{-3} ~5.0×10^{-2}	21.2~15.48	—	—
Chien and Ma (1958)	5.88×10^{-2}	19.1~32.70	—	—
Fei (1981)	1.66×10^{-1} ~4.7×10^{-3}	25.6~22.20	—	—

Debris Flow of the Chui-Sue River Watershed

The Chui-Sue river watershed is located in the southwestern part of the up-stream Ho-Sho River, along the new trans-island highway. In July 31, 1996, severe debris flow occurred in this area when typhoon Herb struck the central area of Taiwan. Several incidents of debris flow occurred since then, in May, 1998, the Ho-Sho bridge crossing the Chui-Sue River was raised and pushed sideway due to damming up of the debris under the bridge. The rock formation of the watershed was mainly interbedded sandstone and shale of the Tertiary Ho-Sho and Nan-Chuang formations with little or no metamorphic action. The geological conditions in the area were fairly fractured with old landslide in the up-stream area.

The microgeomorphic analysis was performed on the Chui-Sue river watershed, the downstream area of the river is the Shen-Mu area. The purpose of the microgeomorphic study is to identify the triggering area, flow area, and the deposit area in the watershed along the debris flow river. Also through the study, the affected areas along with the amount of material erosion and the amount of material deposition at the downstream area could be determined, and which could be used for the subsequent simulation and verification. For the Chui-Sue river watershed, the 10 m by 10 m digital terrain model was built using the aerial photos of the area before and after the typhoon Herb. With the topography and sunshade map, the boundary of the watershed was determined as shown in Fig.1. By running a comparison of the elevations for each point with the same coordinates, the changes of the geomorphic condition of the area could be established. It was found that riverbed became wider and deeper, and the cross section changed from a V shape into a U shape after the Typhoon struck. The triggering area was identified as the area where the major action was erosion, the flow area was identified as the area where both erosion and deposit actions occurred and was about balanced, and the deposit area was identified as the area where the major action was deposit. Results of the study are as shown in Fig. 2. The affected areas and the corresponding amount of material erosion and deposition could be calculated according to the variation of the elevation, the results were listed in Table 2. The total amount of material eroded away from the watershed was 8156397 m^3, and the total amount of material deposited in the area was 5123500 m^3. Therefore, there was a total loss of material of 3032897 m^3 from the watershed. These data are to be used as input for the potential analysis of the area, and provide a basis for discussion of the potential hazard area.

Parameters for Debris Flow Simulation

The parameters needed for the analysis included the hydrological parameters, the geomorphic parameters, and the debris flow parameters. The hydrological parameters were obtained from analyzing the precipitation record in a nearby area by the Central Weather Bureau during typhoon Herb, while the geomorphic parameters were obtained from the microgeomorphic database. With the hydrological analysis, the inflow hydrograph was obtained and combining with the volume of debris, the time record of debris flow was determined as shown in Fig. 3. This debris flow record was then used in the analysis of flow and deposit processes of debris.

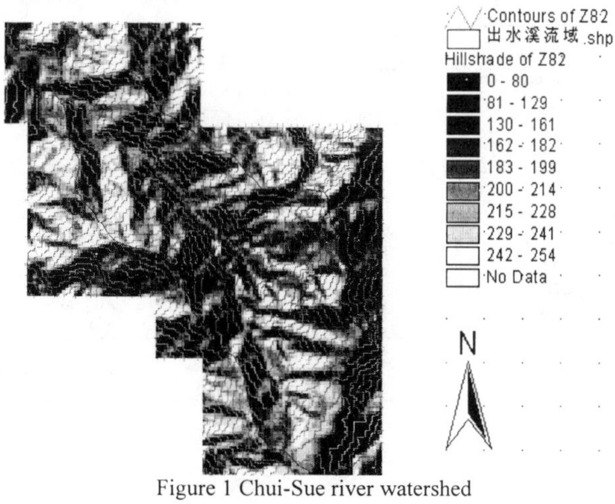

Figure 1 Chui-Sue river watershed

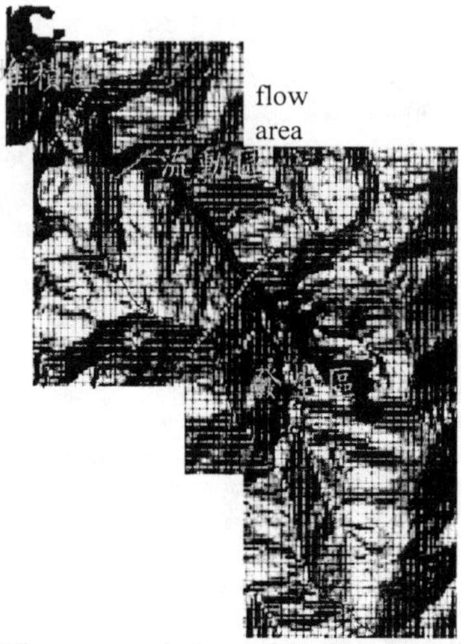

Figure 2 Microgeomorphology of the Chui-Sue river watershed

Table 2 Areas and Corresponding Volumes of Material erosion and Deposition

	Triggering area	Flow area	Deposition area	Landslide area	Total
Volume of deposition (m^3)	833287	922494	3361815	5904	5123500
Projected area of deposition (m^2)	453083	362462	907953	3452	1726950
Surface area of deposition (m^2)	475464	394211	942816	4340	1816831
Volume of erosion (m^3)	5066358	2413641	599183	77215	8156397
Projected area of erosion (m^2)	1064518	626037	180399	18048	1889002
Surface area of erosion (m^2)	1154291	669833	201544	21337	2047005

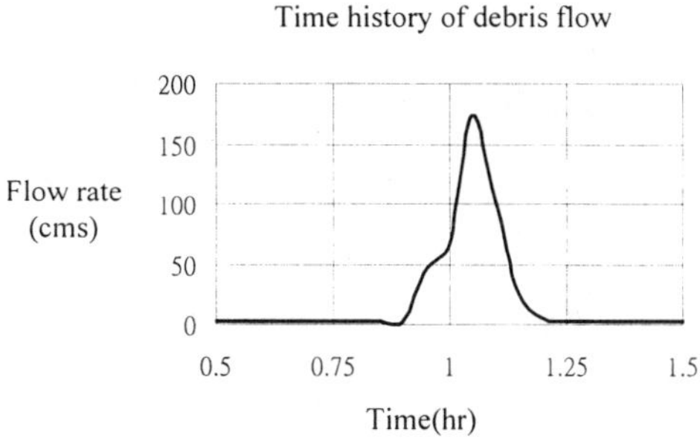

Fig.3 The Time record of debris flow

The debris flow parameters needed contained the mechanical properties and flow properties of the debris material. In order to perform analysis of the flow and deposit process, studies of the mechanical properties of the watershed were needed. Field investigation of the Shen-Mu area was conducted. According to Johnson and Rodine (1984), the grain size distribution of debris flow material at the source area of flow and the material deposited at the deposit area are similar. Therefore, the material at the deposit area was used in both field and laboratory studies. A test pit with the volume of 0.63 m^3 was dug and grain size analysis and determination of the unit weight of the material was performed. Excavated material with grain size smaller than 10cm was brought back for physical and mechanical property test in laboratory. The in-situ moisture content was 5.5%, the dry unit weight was 19.2 kN/m^3, and the void ratio was 0.41 for the field material. The specific gravity determined in the laboratory was 2.71, and according to the Unified Soil Classification System, the material was classified as poorly graded silty gravel (GP-GM). For the mechanical properties of the material, direct shear test was performed on reconstituted specimen with scaled-down grain size due to the limitation of the instrument. A modified weight replacement method was used and the maximum grain size used was 3.81 mm. Results of the direct shear test were as shown in Fig. 4, and the cohesion intercept and friction angle of the reconstituted material were 12.0 kPa and 40°, respectively. The resulting Mohr-Coulomb resistance composed part of the yield stress. For the flow properties of the debris, Jang(1997) performed laboratory study with volume concentration C_V of 0-0.45, and shear rate of 0-3000(1/sec). Considering the difference between field condition and laboratory, the values of parameters were modified. The parameters used in the numerical simulation are as listed in Table 3.

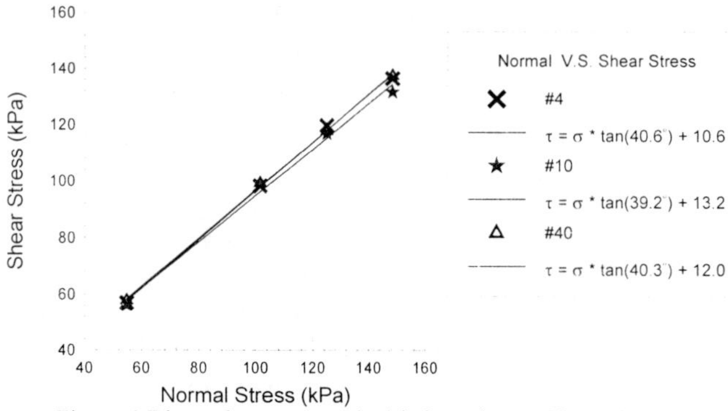

Figure 4 Direct shear test on air-dried specimen (Shen-Mu area)

Table 3 Parameters used in Numerical Simulation

($\eta = \alpha_1 e^{\beta_1 C_v}$ 、 $\tau_y = \alpha_2 e^{\beta_2 C_v}$)

α_1	β_1	α_2	β_2
0.924	14.612	1.2652	16.464

Simulation Results and Verification

Using the 3-D topographical model of the area and with the input parameters discussed in the previous section, the numerical simulation of the debris flow was performed. The average volume concentration of the debris was about 0.5. The resulting deposition conditions are as shown in Fig. 4. The thickness of deposition in the Shen-Mu elementary school was about 4.7 m, and at the location of Ho-Sho Bridge was about 3.48 m. In order to compare the field condition with the numerical results, the average thickness of deposition and erosion of each subarea was computed from the microgeomorphic analysis and was plotted as shown in Fig. 5. Eight grid points along the Chui-Sue River and Ho-Sho River were selected for comparisons, and the locations are as shown in Fig. 6. Comparisons of the thickness of deposition from the simulation and microgeomorphic analysis of the eight grid points are as listed in Table 4. At the grid No. 1969, where the Shen-Mu elementary school located, the results of the numerical simulation were very consistent with the microgeomorphic analysis, and the error is only 3%. For most other locations, the results were not as well consistent. For the study area, the total area of deposition from numerical simulation was about 37% larger than that of the microgeomorphic analysis. Observing the topography map of the area as Fig.5, it was obvious that the inflow of other rivers could have significant effects on the deposition thickness and area of the downstream area. The other two inflow rivers were both debris flow rivers with significant amount of both water flow and debris. The inflow of the rivers could either introduce more debris or flush the deposited material further downstream, which could affect the thickness and area of deposition. It could be verified as shown in Table 4 that downstream locations close to the river inflow area appeared to be more significantly affected than other locations. However, the trend of thickness and area of deposition appeared to be fairly consistent with the microgeomorphic data especially for the upstream area not affected by other inflow rivers.

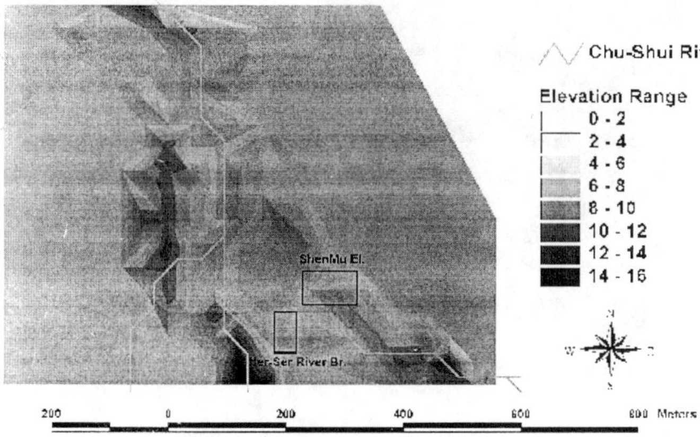

Fig. 4 Results of simulated areas and thickness of debris deposition

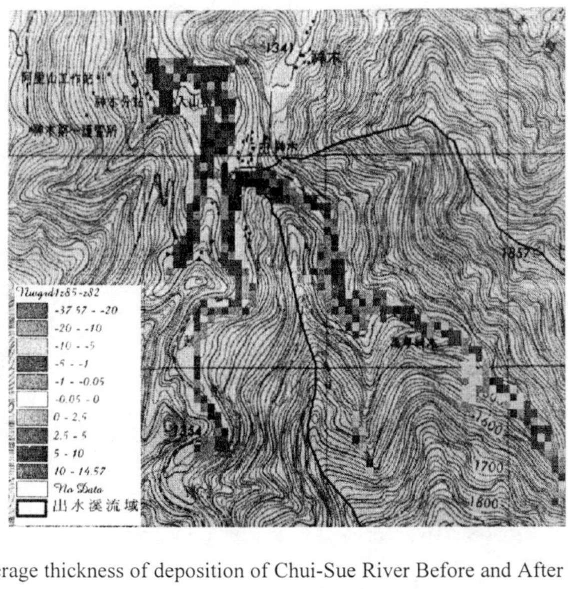

Figure 5 Average thickness of deposition of Chui-Sue River Before and After Typhoon

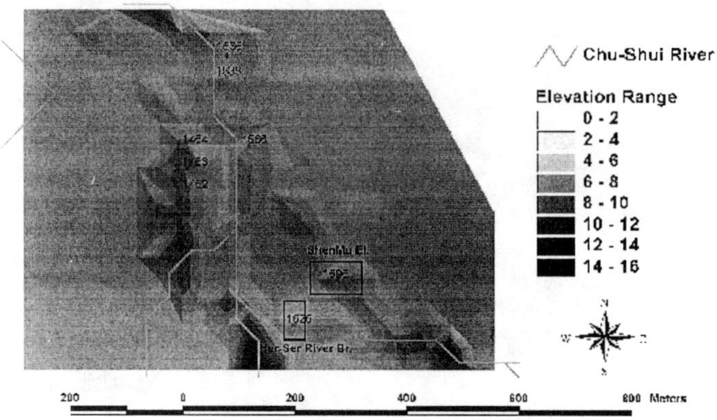

Fig. 6 Locations of grid points for comparison

Table 4 Comparisons of deposition thickness and area

No of grid	Numerical(m)	Micro (m)	Error(%)	Note
1462	7.11	5.58	-27.42	
1463	6.41	6.34	-1.10	
1464	7.82	4.63	-68.90	
1535	3.93	8.57	54.14	
1536	2.14	2.45	12.65	
1566	2.32	8.73	73.42	
1626	3.48	9.02	61.42	Ho-Sho River Bridge
1696	4.71	4.87	3.29	Shen-Mu Elementary School
Deposition area (m^2)	236800	173000	-36.88	

Conclusions

Based on the previous discussions, the following conclusions could be reached:

1. Based on the microgeomorphic analysis performed on the Chui-Sue river watershed, the triggering area, flow area, and the deposit area in the watershed along the debris flow river were identified. The major deposition occurred in the downstream area, and the Chui-Sue river appeared to be a typical valley-type debris flow river.
2. From the microgeomorphic analysis, the total amount of material eroded away from the watershed was 8156397 m3, and the total amount of material deposited in the area

was 5123500 m3, resulting in a net loss of material of 3032897 m3 from the watershed. The amount of debris left in the area was still very high, and which could likely become the source material of further debris flow.

3. In the numerical model, the kinematic term as well as the diffuse term of wave equation were taken into account, and should be able to describe the debris flow appropriately. The simulated results agreed fairly well with the microgeomorphic data for the area not affected by other inflow rivers, and the trends of distribution of the study area appeared to be fairly consistent.
4. The numerical simulation of debris flow provided good information for identifying affected area and its severity. However, the parameters needed in the numerical model were difficult to determine.
5. The parameters for the yield and viscosity stress could have significant effects on the computed results, and yet were difficult to obtain. The laboratory testing conditions were often very different from the field condition, and might resulted in very different values for the parameters. More research into determination of the parameters in the numerical model was needed.

Acknowledgments

This study was a part of the National S&T program for Hazard Mitigation, and was supported by the National Science Council.

References

Jang, C. D., "Teaching material- debris flow", project report of Ministry of Education, 1997.

Johnson, A. M. and Rodine, J. R., "Debris Flow", Slope Instability, John Wiley & Sons Ltd., 1984, pp257-361.

Lin, M. L., et. al., "GIS and characteristics of debris flow rivers of Chen-You-Lan river watershed", NAPHM Annual Report, 1999

Lin, M. L., et. al., "Potential analysis and risk assessment of debris flow rivers of Chen-You-Lan river watershed", NAPHM Annual Report, 2000..

Marsal, R. J., " Mechanical properties of rockfill and gravel materials", Proceedings of the 7th International Conference on Soil Mechanics and Foundation Engineering, Vol.3, 1969, pp.499-506.

O'Brien, J. S. and Julien, P. Y., "Physical process of hyperconcentrated sediment flows", Proceedings of the ASCE Specialty Conference on the Delineation of Landslides, Floods, and Debris Flow, 1985, pp.260-279.

Williams, G. P. and Guy, H. P., "Debris avalanches- a geomorphic hazard", *Environmental Geomorphology, SUNY, Binghamton, N.Y.*, pp25-27.

Uncertainties in Flood Disaster Management

Shou-shan Fan, Ph.D.*

Abstract

Flood disasters caused by extreme hydrologic events have long been a major threat to human well-being. Our struggles against disasters have been documented in historical records since Biblical times. Today, the potential losses due to a major flood disaster are tremendous. As the methods to counter such disasters increase, the public tolerance for such disasters decrease.

In spite of great achievements by scientists, flood disaster management is still not an exact science and involves a great deal of uncertainty. There are several types of uncertainty, including socio-economic, health, and technology induced risk. Due to space limitation, this paper is limited to a discussion of the technology-induced-uncertainty of flood disaster management.

Introduction

Floods can be natural, man-made or a combination of the two. Natural floods can be subdivided into three major types: stream, coastal and urban floods. Man-made floods are products of dam-breaching, war and other human activities. This paper will be limited to a brief discussion of stream floods.

Floods, no matter how violent, need not necessarily cause major damages. Loss of lives can be prevented and economic impact minimized if decision-makers are provided with accurate prior information regarding the major flood characteristics.

In this paper, based on his experiences in chairing an Interagency Computer Model Evaluation Work Group and with the Federal Interagency Committee on Natural Disaster Reduction of the U.S. National Science and Technological Council and elsewhere, the author has conducted an in-depth analysis of the problems and issues and an examination of the deficiencies of the existing mitigation measures. The author hopes that the conclusions of this paper will help reduce the uncertainty in predicting floods and provide impetus for further research.

---

* Chair, ASCE /EWRI Task Committee on Mitigating Floods and Droughts With Modern Technologies
Adjunct Research Professor, University of Wisconsin at Milwaukee

Analysis of Technology-Induced Uncertainty

The major sources of the technology-induced uncertainty in flood disaster management are: complexity of flood characterization, complications induced by human intervention, inadequacy of existing modeling techniques, and most importantly, inadequacy of data needed to calibrate, verify, and implement the heavily data-dependent models or techniques presently available.

2.1 Complexity of Flood Characterization

Floods are complex because they are multi-dimensional and products of three major factors, namely meteorological, physiographic, and human intervention. These three factors are closely interrelated.

a. The meteorological factors which influence flood flows in large rivers are the intensity, amount, and distribution of precipitation.

b. The physiographic factors which influence floods can be subdivided into two major categories: basin and channel characteristics. The basin characteristics are the size and slope of drainage area, permeability, land coverage and uses, and groundwater conditions that directly affect the flood run-off from the basin. The channel characteristics are geometrical and hydrodynamic properties of the channel that control the movement of flood waves.

c. The human intervention factors can be subdivided into two categories: engineering and administration. Major engineering measures involve the construction of dams or reservoirs to regulate the flood flows, the building of levees to confine floods, the diversion of stream-flows to reduce floods, and the forecasting of floods to issue early warnings. Administrative measure involve the urbanization and zoning of flood plains, issuance of flood insurance, etc. Since human intervention factors would significantly affect the physiographic factors previously discussed, and therefore, could complicate the flood characteristics.

Thus, the complexity of flood characteristics makes them very difficult to simulate and predict.

2.2 Uncertainties Due To Inadequacies of Existing Technologies

a. Uncertainty of Major Assumptions and Simplifications

As previously discussed, flood disasters are hard to comprehend. In practice, to make the problems more manageable with the existing techniques, scientists have to make a number of assumptions and simplifications. For example, we normally assume that a complex system can be decomposed into a number of linear subsystems which can be solved with the existing mathematical procedures. In solving the simplified problem with the available point value data, we also need to assume that the flood disaster problems are homogeneous in both time and spatial domains.

When probabilistic or stochastic method is used, one may have to assume that the problems are random in nature.

b. Uncertainty of Safety Design

1. General

The purpose of safety design is to recognize potential hazards of a structure failure and to reduce the risk of failure to an acceptable level. Uncertainty of a structure failure can be minimized if appropriate technologies are applied. Safety design includes two major tasks.

- The selection of an acceptable safety standard
- The design of a structure based on the standard.

2. Uncertainty in Selecting Safety Standard

This is a highly controversial issue. At present, there is no universally accepted standard that has been established in any country. In the United States, two simplified approaches are in use in a rather arbitrary manner. One is based on the philosophy that a structure failure must be prevented at any cost if such failure could cause the loss of human life or catastrophic damage to property. The other is primarily used for evaluating a low or medium hazard failure; it is based on the philosophy that some probable failures could be tolerated.

c. Uncertainty In Selecting An Incoming Flood Prediction Model

1. The Problems

Computer models have been widely used for flood disaster management. While computer models are highly useful tools they can also be a source of misinformation for users who do not understand all the assumptions, capabilities, and limitations of a particular model.

In addition, computer model remains a highly subjective science, involving uncertainties and risks. It depends heavily upon professional judgements for a successful outcome.

Presently, there are two frequently used stream flood prediction techniques: probable maximum flood (PMF) approach and frequency analysis. Each of these methods has its advantages and disadvantages. Each requires different types and amount of data. More importantly, each usually results in different degrees of accuracy.

2. PMF Approach

This method has been widely adopted by many engineers in the United States for the design floods of high hazards projects. It is based on a linear "flood hydrograph theory" and is practically derived by applying the regional probable maximum storm (PMS) values to a pre-selected unit hydrograh. The PMS values can be derived from the local probable maximum precipitation (PMP) values that can be estimated in several ways: generalized PMP charts, storm transposition, statistical methods, and storm models..

In general, the PMF approach is often called deterministic. In reality, it is semi-empirical because it is heavily data dependent. As a result, this method will not produce a unique outcome.

The approach assumes that the PMF is proportional to the PMP value and stresses only the safety issues without considering the other factors, such as economics and environmental issues. However, it is important to note that a structure can fail because of many reasons and that designing a structure to withstand the PMF does not totally eliminate the risk from failure. On the contrary, the PMF may lull the public downstream from the structure into an illusion of absolute safety.

3. Frequency Analysis

Frequency analysis is a powerful technique for estimating inflow design flood that can maximize the project's net benefits because it can incorporate the factors of safety, environment and economics into the design process. Many countries other than the United States, use this approach to estimate their spillway design flood.

The major types of frequency analysis are statistical analysis and probability distribution fitting. Using frequency analysis, one needs to make at least two assumptions: (1) natural phenomena, such as flood and rainfall, are random and amenable to statistical interpretation and probabilistic analysis; (2) the data used are homogeneous in both spatial and time domains.

4. Case Studies

The inadequacies of the existing flood estimation methods can be best demonstrated by two case studies in the United States. One study is the probabilistic analysis of the annual peak floods of the Pecos River near Comstock, Texas, for the 53-year period from 1901 through 1954. In 1954, the Pecos River had a flood that was 8 to 9 times higher than that were previously recorded. Evidently, the analysis had failed to predict the river's 1954 flood.

The second study is the PMF derivation for the northeastern United States (FEMA, 1990). The National Weather Service (NWS) in its 1956 Hydrometeorological Report (HMR) No. 33, estimated the probable maximum precipitation (PMP) used for deriving PMF was

19.2 inches for a 24-hour duration in a 200 square-mile basin of the Deerfield River, a tributary of the Connecticut River in the northeastern United States. Applying the PMP to the Corps of Engineers HEC-1 computer program, the PMF was estimated to be 248,700 cubic feet per second (cfs). In 1978, in its HMR-51, the NWS upgraded its PMP to 21.5 inches. As a result, the PMF was increased to 277,500 cfs.

d. Uncertainty of Routing Floods through Reservoirs

When routing through one or more reservoirs, floods will be modified significantly. The estimation of the floods passing a multiple system of reservoirs, which are located either in series or parallel, or both, in the same basin, should be performed systematically from the most upstream reservoir toward the downstream reservoirs. Also, in a multi-reservoir system, the failure of an upstream reservoir may induce an unexpected sudden increase in flood flow and siltation to the next reservoir downstream. This, in turn, may trigger a chain reaction of failure of one or more of the reservoirs downstream. Such an event would then have the potential to produce much larger amount of damage.

e. Uncertainty of Evaluating Flood Hazard

1. In evaluating a flood hazard due to a structure failure, the following items should be considered.

   - The ability of the upstream structure to withstand potential overtopping.
   - The amount of structural and property damage that would result from complete or partial failure.
   - The downstream loss of life and property that might result from a dam failure.
   - The possible loss of revenues during the rehabilitation of any failure.
   - The incremental cost and benefit at a greater level of safety.

2. Uncertainty Due To Inadequacy of Existing Hazard Evaluation Methods

   Currently, there are three major hazard evaluation methods: visual inspection, numerical inundation analysis, or both.

   Visual inspection is cheap and easy to carry out. However, it is often not dependable because: (1) the site to be inspected could be inaccessible during flood times; (2) the problem area may be invisible to an inspector because it may be underwater; (3) flood may not be at its peak when one inspects the site, and (4) in case of emergency, the visual inspection may not be able to provide adequate information in time for making major decisions (evacuation or others).

   The numerical inundation analysis is accomplished by routing some pre-selected floods from a reservoir through downstream. In practice, dam breaching flood is determined by a hypothetical dam-failure outflow hydrograph and some presumed characteristics of the stream channel downstream.

When a structure suddenly fails, a high speed flood wave moves the sediments accumulated behind the dam into a floodplain downstream. The flood may wash away homes and destroy other properties. In rural areas, crops and livestock are frequently destroyed by flood. Therefore, the flood usually carries a large amount of collected sediment and debris as it sweeps the channel. Also, flooding may cause erosion in some areas and siltation in others.

Upstream from the dam, flood hazard may sometimes become dangerous when the fast rising backwater inundates densely populated areas. Also, the delta formation caused by reservoir sedimentation may raise the nearby groundwater table during floods.

2.3 Uncertainty Due To A Lack of Adequate Data

a. The Need For Data

Data is an essential part of our life. To the modelers, data is critical in developing models. To the model users, data is absolutely necessary for calibration, verification, and utilization of models.

Unfortunately, in the field of flood disaster reduction, the data we need are often nonexistent or not readily available. Therefore, we usually have to collect a great deal of field data. However, data collection can be complex, expensive, and time-consuming. Sometimes, for a small project, it may not be economically justified.

b. Classification of Data Problems

Data adequacy problems are of three major types: quantity, quality, or both. Data quantity and data quality problems are closely related and are usually treated together.

Alternatively, data problems can be classified into four major categories: monitoring and sampling; processing and errors treatment, transmission and dissemination, analysis and interpretation; presentation; and database management. Due to space limitation, only data monitoring problems will be briefly discussed in the section below.

c. Selected Case Studies: Uncertainty of Data Monitoring

In designing a data monitoring program, the following questions must be addressed:

1. What parameters should be monitored?
2. What kind of gages or monitoring instruments should be used?
3. Where should the gaging station be located and how many gages do we need at each station?

4. How many stations are needed, how should they be distributed, and how should they be operated?

The answers to these questions are not easy to determine, and they depend primarily on how well we understand the characteristics of the problem of concern. Other critical factors are local environment, economics, and budget.

As previously discussed, we may not understand the characteristics of the problems and may not know the true values of the parameters to be measured. Therefore, we often may not know how to select the parameters that could describe the problems we study or how to accurately measure the parameters once they are selected.

For example, precipitation, flood flow, channel roughness and channel geometry are four most important parameters in stream flood studies. At present, however, there is no practical way to account accurately for the true value of these four parameters.

Although we are able to measure the geometry of a fixed-bed channel, it is very important to recognize that in an alluvial channel, channel geometry is random in both the spatial and the time domains and interacts continuously with random floods. As a result, the best we can do is to approximate a time and a spatial domain invariable average through indirect ways, using known values of channel geometry and flow characteristics.

Conclusions

1. Flood disaster management problems are site specific, multi-disciplinary, time-varying, and very difficult to solve. Accurate and timely predictions of flood characteristics are critically important to minimize or perhaps even totally eliminate the consequent disastrous losses in property and human life.

2. Flood prediction involves uncertainties. The problems are real and very serious. For example:

   (a) We are presently unable to predict accurately a flood disaster either analytically or statistically.
   (b) Predictions made using different approaches, or even the same techniques applied by different individuals, can yield significantly different results.
   (c) A Congressional report stated that Federal Emergency Management Agency had originally budgeted in 1993 only $292 millions for its entire nation's annual flood relief fund, but the agency spent $2 billion for mitigating a Midwest flood relief alone in that year.

3. It is the author's belief that the uncertainties in flood disaster management are due to one or more of the following factors.

   - We do not thoroughly understand flood characteristics and their underlying physics, and we may oversimplify the problems.

- We may misuse the available mathematical procedures and techniques that are valid only for a point-value problem to solve a larger regional flood problem.

- Human interventions (dam-building, urbanization, floodplain encroachment, etc.) may sometime invalidate the major assumptions underlying the methodologies.

- We also lack adequate data to calibrate and to verify such heavily data-dependent problems.

- At present, the associated uncertainties in data monitoring and modeling cannot be totally resolved, and we can only expect to make a prediction that is within a socially acceptable statistical bound.

4. To effect a meaningful reduction in risk, we should implement the followings.

   (a) Conduct more detailed examination of models and methodologies used by governmental agencies in predicting the characteristics of flood flows and the consequent flood damages on both a regional and a national scale and require a detailed accounting of the elements of uncertainty associated with the prediction.

   (b) Employ modern technologies to improve field data monitoring and to timely disseminate adequate information for the issuance of warnings, the prompt evacuation of people and other appropriate responses.

   (c) Bridge the technical and administrative gaps among individual disciplines, between theoreticians and practitioners, among offices, institutions, and nations. These gaps have made us overextend our limited research funds and manpower resources and have made a comprehensive in-depth research very difficult to carry out.

References

Fan, Shou-shan; (1993), "A Revisit To Data Problems and Databases," Proceedings, International Water Power Conference, Nashville, TN.

Fan, Shou-shan; (1995), "An Examination of Technological Risk and Uncertainty in Natural Disaster Reduction," Proceedings, White House Conference on Risk And Decision Making On Natural Hazards, National Science and Technology Council, Conference Center, Washington, DC.

Fan, Shou-shan; (1999), "Flood Characterization and Lessons Learned," International Water Resources Engineering Conference, American Society of Civil Engineers, Seattle, WA.

Fan, Shou-shan (2001), "Problems of Model Selection in Regulatory Analysis," Proceedings, ASCE-World Water & Environmental Resources Congress, Orlando, Florida, USA

Federal Emergency Management Agency, (1990), Proceedings of the PMP and PMF Workshop, Berkeley Springs, West Virginia.

Rainfall Analysis of Typhoon TRAMI, TORAJI and NARI in Year 2001

Yen-Ming Chang[1], Ruey-Hsing Lu[2] and Shun-Chu Chang[3]

Abstract

There are 9 typhoons that have struck Taiwan in 2001. Among them, Trami, Toraji and Nari separately have caused a lot of damage in the southern, central and northern parts of Taiwan. The rainfalls of these typhoons are analyzed, and their different specific characteristics are described: Typhoon Trami is concentrated in the Kaohsiung metropolitan area; Toraji's rainfall has a very short duration; and Nari has broken rainfall records in many places.

Keywords ∶ Rainfall analysis, Typhoon Trami, Typhoon Toraji, Typhoon Nari, typhoons in year 2001, return period

Introduction

There are 9 typhoons that have struck Taiwan in 2001. Among them, Trami, Toraji and Nari separately have caused a lot of damage in southern, central and northern parts of Taiwan. And these typhoons separately have resulted in 5, 214, 104 people missing or dead. Typhoon Nari especially has caused the largest flooding in northern Taiwan in the recent century. These typhoons are different from "traditional" typhoons, and they have their own specific characteristics. Their rainfall data are analyzed in the following paragraphs for reference. The routes of these typhoons are illustrated in figures 1 to 3.

1. Section chief of water resource division, WCA, MOEA
2. Water resource division, WCA, MOEA
3. Associate engineer, water resource division, WCA, MOEA

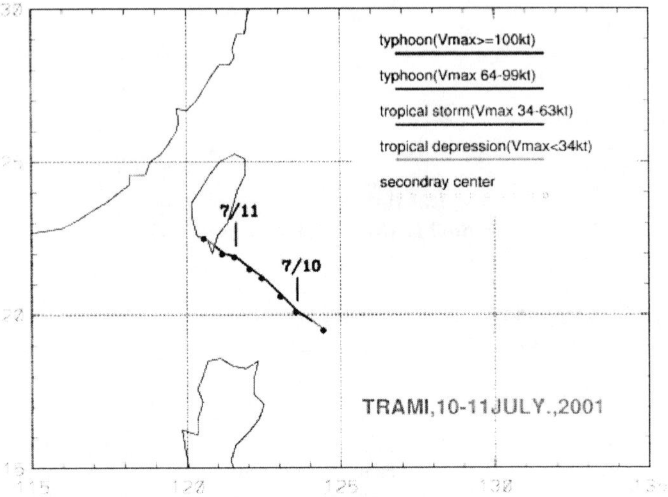

Figure 1: Route of Typhoon Trami

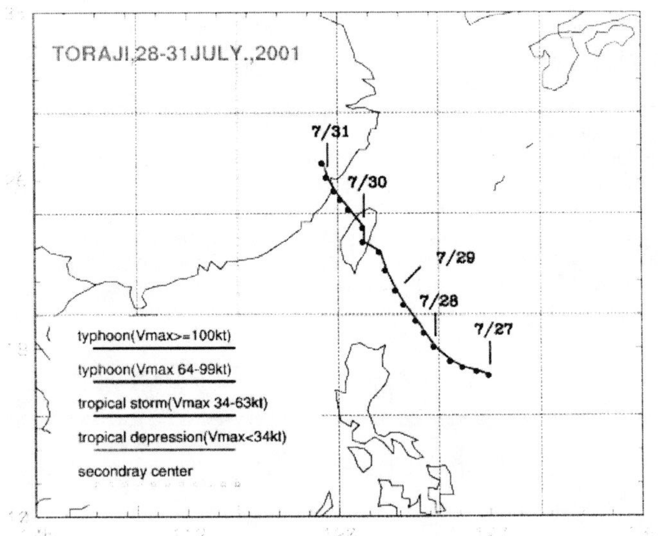

Figure 2: Route of Typhoon Toraji

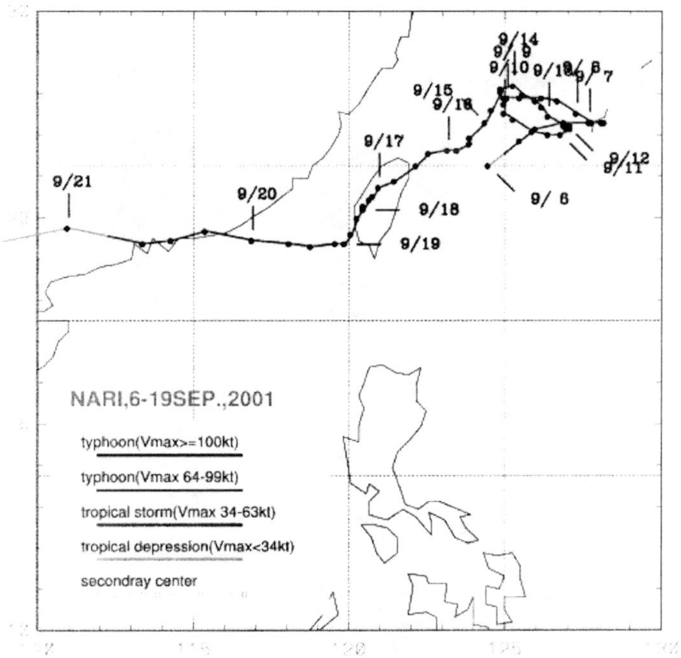

Figure 3: Route of Typhoon Nari

Rainfall analysis

1.Data sources
Rainfall data have been selected from the Central Weather Bureau (CWB)'s 22 man-operated stations and more than 320 real-time rainfall stations(without manpower)and from the Water Conservancy Agency (WCA) 's more than 150 stations. For some example rainfall-station locations, refer to figure 4.

2.Considering the data of record years, frequency analysis (Log-Pearson type 3 distribution) has used data from CWB's 22 man-operated stations and WCA's 150 stations. On the other hand, isohyets have been mainly produced by the data of the CWB's real-time rainfall stations.

3.Rainfall data and analysis of return periods are listed in tables 1 and 2 (CWB stations) and tables 3 and 4 (WCA stations) . Due to the limited allotted pages for this report and to the fact that only a few stations have had heavy rainfall, data of Typhoon Trami have been omitted. We have found the following facts:

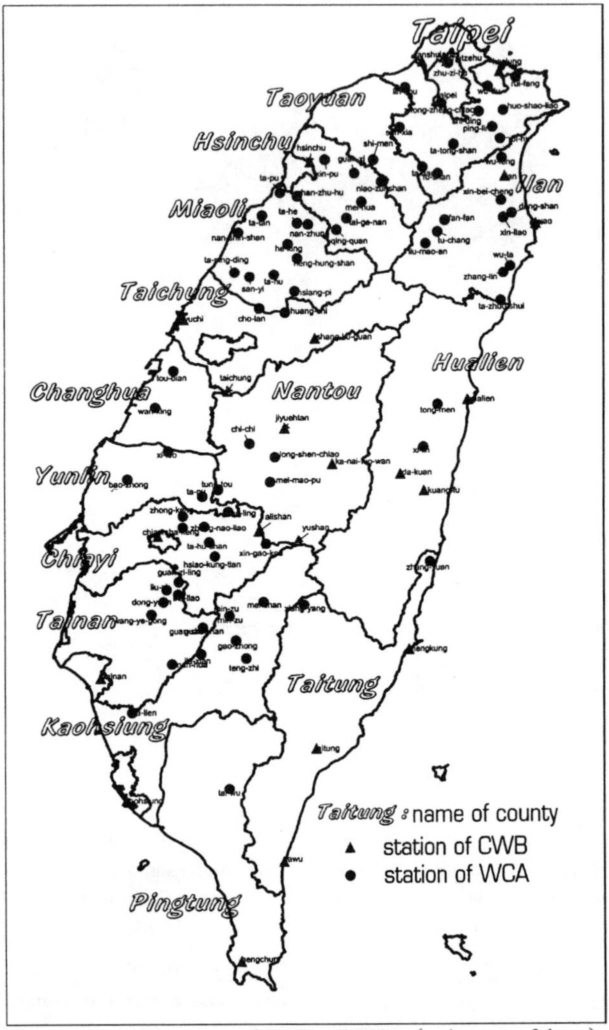

Figure 4: Rainfall stations of CWB and WCA（only parts of them）

Table 1: Rainfall of Typhoon Toraji and analysis of return period (CWB)

	Date	7/29	7/30	7/31	1-day rainfall return period(years)	2-day rainfall return period(years)	3-day rainfall return period(years)
Station	hour	0~24	0~24	0~24	7/30		
1	Keelung	2.0	22.0	0.0	<1.11	—	—
2	Ilan	18.0	15.0	0.0	<1.11	—	—
3	Suao	34.0	42.0	0.0	<1.11	—	—
4	Anpu	8.0	47.0	T	<1.11	—	—
5	Yangmingshan	9.0	41.0	0.0	<1.11	—	—
6	Tamsui	1.0	30.0	0.5	<1.11	—	—
7	Taipei	2.0	33.0	0.0	<1.11	—	—
8	Hisnchu	0.0	206.0	1.0	4.3	2.0	<2
9	Taichung	0.0	230.0	0.1	4.4	2.0	<2
10	Wuchi	0.0	455.0	8.0	260.0	28.0	21.0
11	Jiyuehtan	6.0	333.0	0.0	6.7	3.8	2.8
12	Yushan	49.0	400.0	0.0	50.0	4.2	2.1
13	Alishan	43.0	715.0	0.0	10.5	3.9	2.8
14	Chiayi	1.0	287.0	11.0	9.5	3.9	2.6
15	Tainan	4.0	21.0	0.8	<1.11	—	—
16	Kaohsiung	0.5	3.0	5.0	<1.11	—	—
17	Hualien	31.0	180.0	0.0	<2	—	—
18	Chengkung	22.0	16.0	0.0	<1.11	—	—
19	Taitung	18.0	3.0	0.0	<1.11	—	—
20	Tawu	9.0	25.0	0.0	<1.11	—	—
21	Hengchun	11.0	3.0	0.0	<1.11	—	—
22	Penghu	0.0	2.0	T	<1.11	—	—

A. Rainfall of Typhoon Trami is concentrated in the Kaohsiung area (refer to figure 5). The 1-day rainfall return period is 50 years. But when referring to table 5, the intensities of Typhoon Trami show to be more intensive than before. Duration longer than 6 hours, the return periods are longer than 50 years. Duration longer than 10 hours, the return periods are almost longer than 100 years. The design capacity of the drainage system of Kaohsiung's metropolitan area was insufficient to handle the discharge. Excessive rainfall flooded into the basement of buildings, emerged cars and power supplies, and caused severe damage.

B. Typhoon Toraji is of another type. The maximum 1-day rainfall is 715mm at Alishan (table 1). Owing to Alishan being in a mountainous area, the return period of this rainfall is only 10 years. The maximum 1-day rainfall return period is 260 years at Wuchi, but the 1-day rainfall at Wuchi is only 455mm (located at the coastal area). Next Wuchi is Lung-shen-chiao of WCA. (Referring to table 3, only those stations with rainfall amounts larger than 350mm has been picked out.) Its 1-day rainfall return period is 98 years, but the 1-day rainfall is only 513mm. Referring to table 6, the rainfall intensities and their return periods of several different districts are listed. The maximum 1-hour rainfall is 146.5mm at Kuang-fu; it has a 251 years return period. The maximum 3-hour rainfall is 390mm also at Kuang-fu; it has a 392 years return period. Table 6 shows that the return periods of shorter durations （1~3 hours） are larger than those of longer durations. According to the different durations of the isohyets（refer to figures 6 and 7）, they illustrate the rainfall concentrated in the mountainous areas of Hualien county and between Miaoli county and Chiayi county of the western part of Taiwan. In addition, the centers of the storms are illustrated in figures 8 to 9. Typhoon Toraji is special in that it has a short rainfall duration. At Kuang-fu and Lung-shen-chiao, 80% of the rainfall has occurred in 3 hours. In 6 hours, this figure becomes 90%. Compared to other typhoons with the similar route, Toraji has the shortest duration of rainfall. The debris flow caused by heavy rainfall intensity and the poor vegetation coverage（that was caused by September 21, 1999's earthquake）has destroyed everything and has resulted in severe damage.

Table 2: Rainfall of Typhoon Nari and analysis of return period (CWB)

Date		9/16	9/17	9/18	9/19	1-day rainfall return period (years)	2-day rainfall return period (years)	3-day rainfall return period (years)
Station	hour	0~24	0~24	0~24	0~24			
1	Keelung	270.0	227.0	70.0	1.0	11	54	42
2	Ilan	182.0	321.0	34.0	13.0	9	13	10
3	Suao	173.0	316.0	44.0	13.0	4	5	4
4	Anpu	518.0	288.0	255.0	5.0	8	11	15
5	Yangmingshan	686.0	340.0	246.0	4.0	19	20	23
6	Tamsui	255.0	237.0	153.0	0.0	9	67	99
7	Taipei	246.0	425.0	164.0	T	204	494	788
8	Hisnchu	77.0	335.0	397.0	0.0	143	797	836
9	Taichung	69.0	309.0	96.0	T	12	9	10
10	Wuchi	5.0	371.0	143.0	0.0	70	46	39
11	Jiyuehtan	10.0	50.0	9.0	T	<1.1	<1.1	<1.1
12	Yushan	17.0	219.0	46.0	10.0	1.7	1.3	1.2
13	Alishan	17.0	319.0	78.0	0.5	1.6	1.4	1.3
14	Chiayi	3.0	288.0	775.0	T	700	722	447
15	Tainan	T	15.0	226.0	4.0	3	1.8	1.6
16	Kaohsiung	0.0	4.0	179.0	71.0	1.9	1.9	1.6
17	Hualien	1.0	68.0	71.0	20.0	<1.1	<1.1	<1.1
18	Chengkung	0.0	1.0	18.0	37.0	<1.1	<1.1	<1.1
19	Taitung	0.0	2.0	1.0	64.0	<1.1	<1.1	<1.1
20	Tawu	0.0	2.0	5.0	51.0	<1.1	<1.1	<1.1
21	Hengchun	0.0	1.0	23.0	104.0	<1.1	<1.1	<1.1
22	Penghu	0.0	15.0	206.0	52.0	7	6	5

Table 3: Rainfall of Typhoon Toraji and analysis of return period (WCA)

Date		7/29	7/30	7/31	1-day rainfall return period(years)	2-day rainfall return period(years	3-day rainfall return period(years)
Station	hour	0～24	0～24	0～24	7/30		
2nd RBMB Hsinchu Miaoli Area	Heng-lung-shan	1	370	0	4.9	2.9	2.3
	Ta-hu	0	366	0	5.4	2.5	2.1
	Nan-shih-shan	1	395	3	44.6	27.3	19.4
3rd RBMB Taichung Area	Hsiang-pi	0	383	0	4.8	2.7	<2
	Cho-lan	2	365	0	12.1	4.3	3.2
4th RBMB Changhua Nantou Area	Nei-mao-pu	8	463	0	21.3	8.8	8.3
	Lung-shen-chiao	9	513	0	98.3	23	19.7
	Chi-chi	0	350	0	11.9	4.7	3.6
	Tsao-ling	5	401	0	4.6	2.2	<2
	Tung-tou	0	401	0	11.7	5	3.7
5th RBMB Chiayi Area	Ta-pu	1	352	0	9.3	4.5	3.2
	Ta-hu-shan	16	497	0	6	3.4	2.5
	Hsiao-kung-tian	12	412	0	3.6	<2	<2
6th RBMB Tainan Area	Guan-shan	10	433	5	4.4	2.2	<2
7th RBMB Kaoshiung Pingtung Area	Mei-shan	34	556	0	16.3	4.2	2.5
	Gao-zhong	16	457	5	5.5	<2	<2
	Teng-zhi	26	463	3	6.7	<2	<2
	Min-zu	18	466	2	12.7	2.9	<2
	Tai-wu	216	207	12	<2	<2	<2
	Xin-gao-kou	45	578	0	18.4	5.5	4.5
8th RBMB Taitung Area	Zhang-yuan	234	118	0	<2	2.3	<2
	Xiang-yang	35	406	0	6.6	2.5	<2
9th RBMB Hualien Area	Xi-lin	103	446	0	7.4	4.2	3.1
	Tong-men	84	295	0	2.9	2.2	<2

RBMB: River Basin Management Bureau, a subordinate organization of WCA.

Table 4: Rainfall of Typhoon Nari and analysis of return period （WCA）

Date / Station	hour	9/16 0～24	9/17 0～24	9/18 0～24	1-day rainfall return period(years)	2-day rainfall return period(years)	3-day rainfall return period(years
1st RBMB Ilan Area	Wu-feng	437	352	42	39	353	239
	Liu-mao-an	106	820	117	218	147	108
	Tu-chang	94	1011	290	576	291	178
	Fan-fan	128	779	498	29	53	41
2nd RBMB Hsinchu Miaoli Area	Lin-kou	277	363	232	714	1563	2122
	Guan-xi	150	319	241	11	43	104
	Xin-pu	110	308	351	47	842	1183
	Tai-ge-nan	381	270	61	9	40	47
	Niao-zui-shan	593	327	114	11	67	100
	Ta-pu	55	351	407	126	704	696
	Nan-chuang	234	374	115	16	56	148
	Shan-zhu-hu	77	334	310	21	130	164
	Ta-he	104	383	177	13	24	33
	Ta-tan	41	277	353	39	370	442
	Nan-shih-shan	28	320	273	20	182	187
	Ta-ping-ding	52	600	222	126	141	164
4th RBMB Changhua Nantou Area	Tou-bian	6	424	168	351	289	191
	Wan-xing	13	395	211	134	247	229
	Xi-luo	5	318	254	14	58	47
	Tung-tou	7	449	269	17	38	33
5th RBMB Chiayi Area	Bao-zhong	12	193	320	76	605	185
	Ta-pu	6	553	289	35	69	76
	Zhong-keng	3	358	361	20	305	257
	Zhang-nao-liao	2	611	634	15	90	92
	Sha-keng	2	410	422	40	278	303
	Ta-hu-shan	0	561	697	17	79	59
	Hsiao-kung-tian	0	528	601	13	781	452
	Guan-zi-ling	0	258	756	267	317	237
	Liu-xi	0	176	756	219	252	143
	Bei-liao	0	285	797	335	268	227
	Dong-yuan	0	150	718	205	249	128
6th RBMB Tainan Area	Wang-ye-gong	0	93	597	859	551	552
10th RBMB Taipei Keelung Area	Shi-men	162	324	254	26	239	403
	San-xia	198	269	392	388	466	674
	Ta-bao	371	440	249	37	416	1361
	Ping-lin	602	393	169	32	844	906
	Pi-hu	600	409	116	183	1100	895
	Fu-shan	563	475	138	11	234	392
	Ta-tong-shan	507	436	185	72	1254	851
	Shi-ding	491	510	133	57	793	854
	Zhong-zheng-chiao	212	466	240	267	539	832
	Wu-du	447	441	95	14	34	26
	Huo-shao-liao	564	358	178	36	79	83

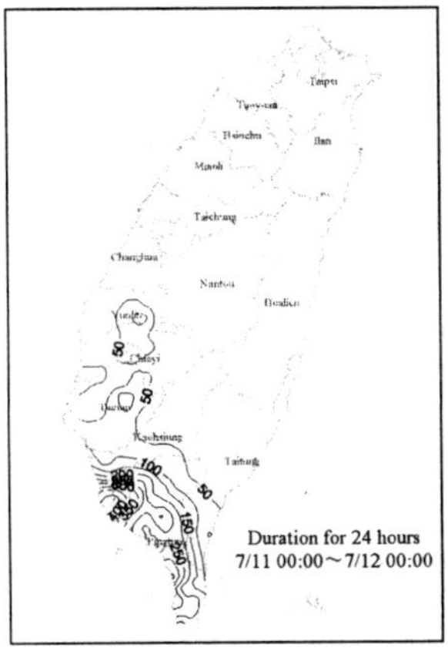

Figure 5. Isohyets for 24-hour duration of Typhoon Trami

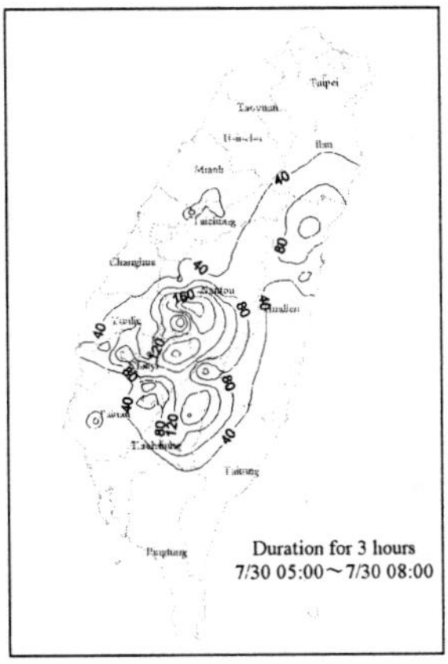

Figure 6. Isohyets for 3-hour duration of Typhoon Toraji

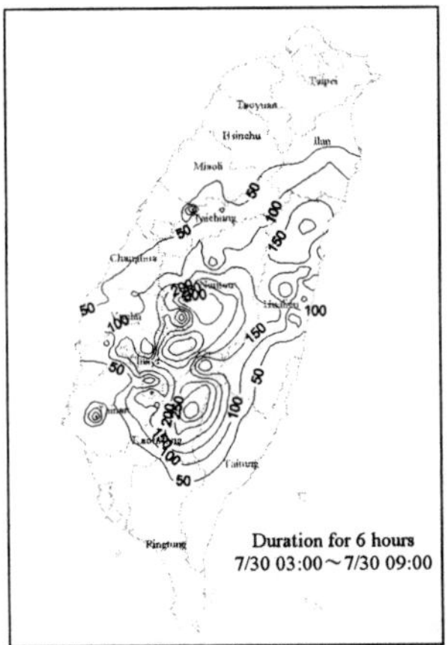

Figure 7. Isohyets for 6-hour duration of Typhoon Toraji

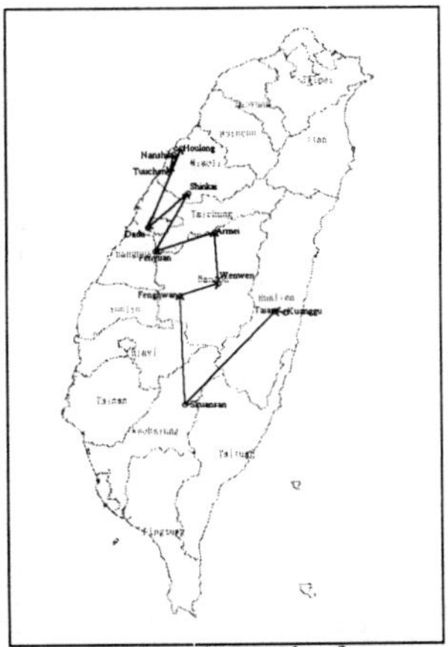

Figure 8. Moving path of storm center for 1-hour duration of Typhoon Toraji

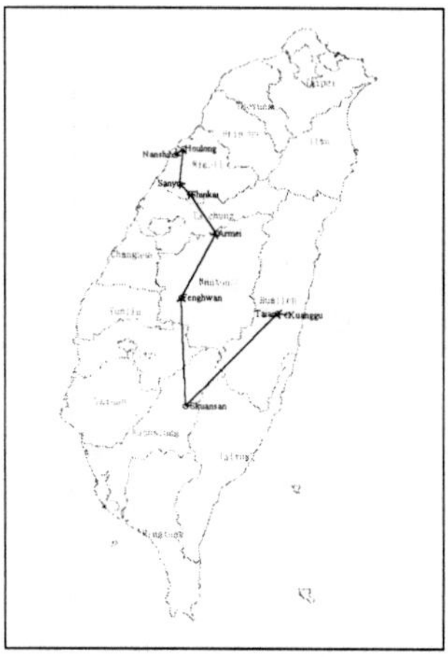

Figure 9. Moving path of storm center for 3-hour duration of Typhoon Toraji

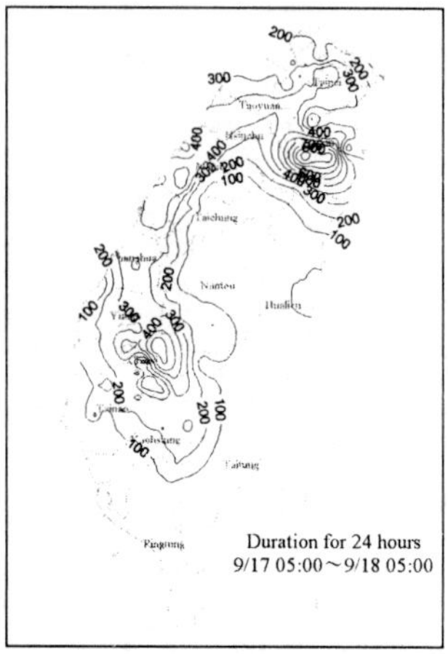

Figure 10. Isohyets for 24-hour duration of Typhoon Nari

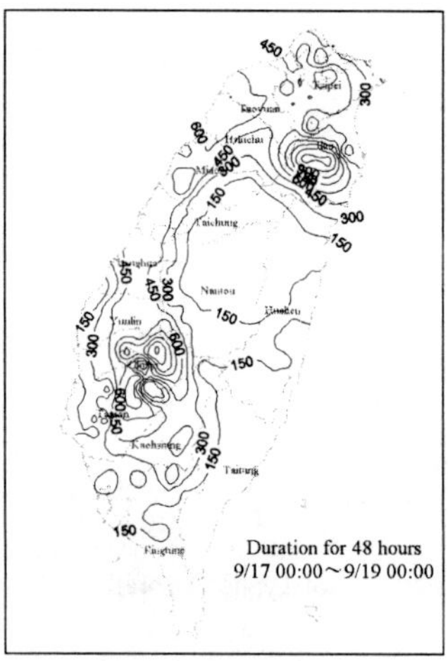

Figure 11. Isohyets for 48-hour duration of Typhoon Nari

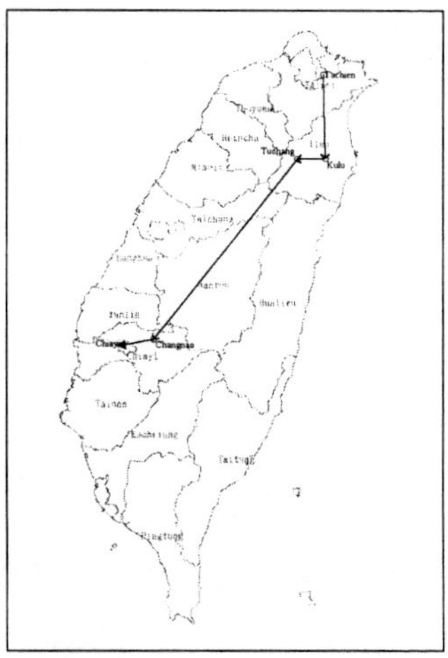

Figure 12. Moving path of storm center for 24-hour duration of Typhoon Nari

Table 5: Rainfall intensities and return periods of Kaohsiung's metropolitan area

Station	Kaohsiung		Tsaoying		Fengshan	
Duration (hour)	Intensity (mm/h)	Return period (years)	Intensity (mm/h)	Return period (years)	Intensity (mm/h)	Return period (years)
1	119.5	36	126.5	46	92	11
2	92.8	28	116.8	73	86.5	22
3	83.7	26	109.5	96	87.3	41
4	76.3	41	87.6	67	81.5	49
5	69.6	45	76.6	63	74.3	55
6	70.1	68	71.2	72	68.8	63
8	59.3	74	64.8	99	60.8	81
10	51.8	83	54.9	99	54.3	96
12	44.5	82	46.5	95	47.3	101
24	22.9	85	24.2	113	24.7	123

C. Typhoon Nari is the weirdest typhoon in the history of Taiwan. Its route from northeast to southwest is different from all other typhoons. This typhoon stayed in Taiwan for over 49 hours, longer than any previous typhoon. Referring to tables 2 and 4 (in this table, only rainfall amounts larger than 500mm and return periods longer than 30 years are listed), the amount of rainfall is astonishing! At least 18 rainfall stations have broken their historic records. Some stations (for example, Lin-kou's 3-day rainfall and Wang-ye-gong's 1-day rainfall) even have twice the original maximum data! The maximum 1-day rainfall is 1,011mm at Tu-chang. This is tragic enough, but a continuous heavy rainfall was the horrific factor that has caused a lot of disaster. From tables 2, 4 and 7, in the northern part of Taiwan, rainfall occurred on September 16th to September 18th. But at the western plain in the middle of Taiwan, rainfall concentrated on September 17th and September 18th. Referring to isohyets of figures 10 and 11, two heavy rainfall centers have occurred. Figure 12 shows that Nari cut a swath through the western plain of Taiwan, from Taipei to Tainan, and skipped only the Taichung area.

Table 6: Rainfall intensities and return periods of Typhoon Toraji

Area	Station		1 hr	2 hrs	3 hrs	6 hrs	12 hrs	18 hrs	24 hrs	Remark
Miaoli	Ta-hu	Intensity (mm/hr)	87.0	79.0	73.7	53.0	29.6	19.9	15.3	
		Return period (years)	30	108	140	51	9	4	3	
	Heng-lung-shan	Intensity (mm/hr)	93.0	74.0	66.3	51.5	29.8	20.5	15.4	
		Return period (years)	74	43	41	25	6	3	2	
Taichung	Shuang-chi	Intensity (mm/hr)	79.0	78.5	66.0	44.3	26.5	18.3	13.9	
		Return period (years)	13	192	157	33	6	3	2	
	Shang-ku-Guan (CWB)	Intensity (mm/hr)	87.0	75.8	63.2	45.4	27.8	19.8	15.3	
		Return period (years)	46	108	80	35	7	3	2	Refer to WCB station
Nantou	Ka-nai-tuo-wan (CWB)	Intensity (mm/hr)	89.5	69.0	59.8	49.8	40.9	29.7	23.0	
		Return period (years)	573	538	516	554	565	325	165	Refer to WCB station
	Long-shen-quio	Intensity (mm/hr)	138.0	112.0	102.0	71.7	39.8	28.8	21.8	
		Return period (years)	313	310	371	394	298	276	216	
Kaohsiung	Teng-zhi	Intensity (mm/hr)	95.0	77.0	62.3	47.0	37.5	26.9	20.4	
		Return period (years)	99	100	49	31	30	10	5	
	Min-zu	Intensity (mm/hr)	80.0	76.0	65.7	50.0	38.0	26.7	20.2	
		Return period (years)	7	27	29	28	28	12	7	
Hualien	Da-kuan (CWB)	Intensity (mm/hr)	84.5	81.0	79.3	67.4	42.58	30.5	23.13	
		Return period (years)	4	9	15	16	6	3	2	Refer to WCB station
	Kuang-fu (CWB)	Intensity (mm/hr)	146.5	136.3	130	77.5	40.75	Data missed	Data missed	
		Return period (years)	251	339	392	179	17	Data missed	Data missed	Refer to WCB station

Table 7: Rainfall intensities and return periods of Typhoon Nari

Area	Station		1 hr	2 hrs	3 hrs	6 hrs	12 hrs	18 hrs	24 hrs
Ilan	Fan-fan	Intensity (mm/hr)	127.0	108.0	96.7	77.2	53.3	46.8	39.3
		Return period (years)	35	41	44	45	30	32	24
	Xin-bei-cheng	Intensity (mm/hr)	110.0	93.0	70.3	56.8	34.4	25.7	20.7
		Return period (years)	18	27	19	35	24	20	16
Hsinchu	Niao-zui-shan	Intensity (mm/hr)	73.0	64.5	61.7	52.0	37.8	30.4	24.8
		Return period (years)	4	6	9	10	7	5	4
	Nan-chuang	Intensity (mm/hr)	74.0	59.5	54.0	39.0	24.6	18.8	17.1
		Return period (years)	7	9	12	9	5	4	5
Chiayi	Zhang-nao-liao	Intensity (mm/hr)	95.0	95.0	91.0	72.7	49.5	45.2	37.8
		Return period (years)	5	36	52	52	34	42	32
	Hsiao-kung-tian	Intensity (mm/hr)	94.0	67.0	69.3	54.5	34.3	40.1	33.3
		Return period (years)	16	11	52	54	21	185	135
Tainan	Liu-xi	Intensity (mm/hr)	97.0	95.0	90.3	60.7	47.8	41.8	33.0
		Return period (years)	5	20	43	46	115	175	161
	Nan-hua	Intensity (mm/hr)	53.0	46.0	38.0	33.3	23.6	24.1	19.8
		Return period (years)	<2	<2	<2	8	18	359	368
Tainan	A-lien	Intensity (mm/hr)	67.0	64.0	62.3	41.7	23.6	19.0	15.3
		Return period (years)	3	20	180	114	19	20	15
Taipei	Huo-shao-liao	Intensity (mm/hr)	85.0	65.5	58.7	50.5	38.8	34.7	31.7
		Return period (years)	33	23	25	115	254	345	491
	Wu-du	Intensity (mm/hr)	115.0	103.5	79.0	54.7	49.1	39.9	32.1
		Return period (years)	313	246	137	102	263	359	437
	Ta-bao	Intensity (mm/hr)	58.0	48.0	42.0	34.7	29.8	25.4	22.5
		Return period (years)	3	8	11	29	89	103	107
	Bi-hu	Intensity (mm/hr)	78.0	60.0	50.3	46.7	40.2	35.6	32.8
		Return period (years)	13	12	9	46	295	647	1087

Discussions and conclusions

1.In Taiwan, the longest rainfall data is about 100 years. Most WCA's rainfall stations only have 30, 40 years data. Using these data to analyze, to plan, or to design hydraulic works is obviously insufficient. We must carry on our hydrologic observations. And for safe reasons, if possible, we should use longer return periods for the design standards.

2.In year 2001, these 3 typhoons showed us that the intensities of rainfall were quite larger than before. Is this a special case? Or the hydrologic situation in Taiwan is different from the past decades? We should pay much more attention to the global weather changes and beware of the uncertainty of hydrologic conditions.

3.The design return periods of drainage systems in Taiwan usually are less than 30 years. In rural area and farmlands, this standard meets the need. But in the metropolitan area of Taiwan, the power supplies of buildings are all installed in the basement. Therefore, in the case where floods may enter the basement, the entire building（including tap water supply）is affected. We should consider seriously these situations and solve the problems.

4.Debris flow will become a real nightmare for Taiwan in the next few years（or decades）, especially in the central part of Taiwan. Debris flow deposits a large amount of sand and rocks into the river, also lifts the riverbed up, and changes the cross sections of rivers. We must re-evaluate our flood-prevention plans and reconsider how to mitigate the damage of floods.

Acknowledgements

Thanks to the CWB. Database and FTP server of CWB help a lot. Thanks to Mr. Jin-Shan Huang, commissioner of WCA, he gave us this topic and very useful advice. Also thanks to Mr. Jung-Feng Weng and other colleagues. They collected data and assisted in drawing maps. Also thanks to Dr. Jet-Chau Wen and Miss Resta of National Yunlin University of Science & Technology. They helped to revise this article.

References

WCA, MOEA 「Special report on Typhoon Toraji and its hydrologic analysis. 」Sep. 4 2001（in Chinese）

WCA,TPG 「Statistics and return period analysis of storms in Taiwan」Nov. 1998（in Chinese）

Water Conservancy Bureau, TPG 「Analysis of rainfall intensity and duration in Taiwan area」Jun. 1987（in Chinese）

Watershed Management for Mitigation of Disasters Caused by Extreme Hydrologic and Tectonic Events in Taiwan

Jie-Daer Cheng[1], Huei-Long Wu[2], Jing-Shan Hwang, Cheng-Daw Hsieh[3] and Kenneth N. Brooks[4]

Abstract

As a result of steep topography, young and weak geological formations, loose soils, disastrous floods, landslides and debris flows associated with extreme hydro-meteorological and tectonic events such as heavy typhoon rainfalls and earthquakes often cause significant losses of and damage to homes, human lives and property and pose serious challenges to the sustainable management of land and water resources in Taiwan. Considerable efforts are made for mitigation of these natural disasters, but most often in a crisis management manner that lacks a comprehensive and long-term perspective. Developing effective measures and responses to mitigate these natural disasters is compounded by difficulties in implementation of proper land use planning and regulation due to land scarcity, inadequate policies and lack of cooperation between people and institutions. Problems and solutions of disaster mitigation related to floods and debris flows are discussed in the context of sustainable land and water management in Taiwan. Engineering approaches to modify natural systems are contrasted with approaches that require changing human behavior. Watershed management practices include measures to change the natural system, human behavior, or both, for mitigating disasters related to debris flows and floods. The success of mitigation measures depends not only on characteristics and magnitudes of disasters but also on human responses related to socioeconomic, cultural and political factors.

Introduction

Floods and debris flows claim hundreds of lives and cause millions of dollars of property damage throughout the world each year. As a result of steep topography, young and weak geological formations, frequent seismic shakings, erodible soils, earthquakes, heavy rainfall and improper land uses, Taiwan is also subject to disastrous floods landslide and debris flows that often cause significant losses and damages to human lives as well as public and private properties. These disasters have brought increasing attention of the

1. Professor, Soil & Water Conservation Department and Director, Research Center for Conservation of Water Resources and Disaster Prevention, National Chung-Hsing University, Taichung, Taiwan
2. Director General, Soil and Water Conservation Service, Council of Agriculture, Nantou, Taiwan
3. Director General and Division Chief, Water Resources Bureau, Ministry of Economic Affairs, Taipei, Taiwan
4. Professor, Department of Forest Resources, University of Minnesota, St. Paul, MN, U.S.A.

general public,elected representatives and responsible government agencies alike, leading to the gradual but continuous efforts toward floods and debris flow hazard mitigation programs.

This paper focuses on problems and solutions of floods and debris flows caused by hydrometeorlogic and tectonic extremes in Taiwan. It assesses the basic concepts and approaches related to Taiwan's floods and debris flows hazard mitigation program, which until now still relies very heavily on engineering methods, and to explore possible alternatives and options. Emphasis in this paper is on the mitigation of disasters related to floods and debris flows in Taiwan using an overall watershed management perspective.

Floods and Debris Flows in Taiwan

Few places in the world experienced the problems of hydro-meteorological extremes like Taiwan. The average annual rainfall is 2,500 mm with 3,000 to 4500 mm in some high mountain regions. About 80% of the annual rainfall is concentrated in the May-October typhoon season. Approximately 3 typhoons pass through the island annually and often bring more than 100 mm/hr or as much as 1,000 mm/day of rainfall.

Two-thirds of Taiwan are rugged mountains and hills. Most mountainous areas are very steep, with slopes usually exceeding 45%. About 59 percent of Taiwan's land area are covered with forests. Fifty percent of slopeland crops grow on slopes steeper than 30%. According to slopeland inventory in 1998, there were still 110,000 ha of cultivated land requiring conservation treatments. Nearly all the slopeland soils are subject to serious erosion when exposed. The problem is aggravated by improper cultivation on steep slopes.

Lee et al. (1990) surveyed 40 watersheds (total area of 20,428 km^2) in Taiwan in which 9,900 landslides occurred with a cumulative slide area of 16,172 ha. During the typhoon season, many landslides trigger debris flows in upland drainages. The characteristics and effects of more recent typhoon-caused debris flows for selected areas in Taiwan are summarized in Table 1 and Table 2.

In June 1990 Tungmen debris flow killed 35 people, and another 19 lives were lost in debris flows caused by Typhoon Herb in 1996. Excessive rainfall of high intensity was the main cause of all the above debris flows. A total rainfall of 1,987 mm of was measured over a 42 – hour period at one high elevation location during typhoon Herb (July 31-Aug.1) in 1996 and led to devastating landslides, debris flows and flooding (Table 1). More recently, a largest earthquake of this century in central Taiwan occurred on September 21, 1999 had caused of more than 2,400 human lives and loss of billions dollars worth in houses, buildings, roads, bridges and dams and induced 2,365 landslides and soil mass movement (Table 2).

On the other hand, billions of dollars are spent each year for flood prevention, flood forecasting, and flood recovery in Taiwan — yet floods and flood related disasters occur frequently due to heavy typhoon rainfalls. Most recently, in July and September respectively of this year 2001, heavy rainfalls during two typhoons had caused tremendous damages to lives, properties roads and bridges (Table 3).

Table 1. Selected Debris Flow Events (1990-1996) in Central Taiwan (From Cheng et al. 1997)

Debris Flow Location	Date	Rainfall Event	Impacts on Life and Property
Tung-Men, Hualin	23 June 1990	475 mm / 3 hr	29 deaths, 6 missing, 7 injured, 24 houses destroyed, severe road damage
Er-Bu-Keng, Nantou	31 July - 1 Aug. 1996	>700 mm in less than 2 days	5 deaths, 10 houses & 3.8 ha fruit orchard destroyed
Tung-Fu, Nantou	31 July - 1 Aug. 1996	>1300 mm in less than 2 days	2 deaths, 18 houses destroyed or damaged
Shen-Mu Villiage, Nantou	31 July - 1 Aug. 1996	>1600 mm in less than 2 days	5 deaths, 6 injured, 8 houses destroyed, 3 ha fruit orchard damaged

Table 2. Landslides and Streams with Debris Flow Hazards in Central Taiwan after the 7.2 Ritcher Scale Earthquake on September 21, 1999

Landslides		Number of streams with debris flow hazard
Number	Areas (ha)	
25,845	15,977	377

Table 3. Features and Impacts of Most Recent Floods and Debris Flows Occurred This Year (2001) in Taiwan

Location	Date	Features + Causative Rainstorms	Impacts
Hualien County (Kuang-Fu.Feng-Lin Area)	July 31-Aug 1 2001 Typhoon Toraji	a. Debris Flows, Muddy Floods with Large Boulders b. (490-700mm/3h).	Over 60 Deaths and Missing.
Nantou County (Hsin-Yi,Shu-Li,Lu-gu, Chu-shan Area)		a. Muddy Floods with Large Boulders, Debris Flows b. 600-1000mm Total Rainfall.	Over 100 Deaths and Missing. Significant Damages to dikes, roads, bridges+ buildings.
Tapei City, Northern + Central Taiwan	Sept.15-18 2001 Typhoon Narry	a. Mainly Floods plus Debris Flows at Various Locations b. 425 mm/day (200 yr event)(Taipei). 835 mm/3day (over 500 yr event).	Over 80 Deaths and Missing. > 2000 ha flooded.

a. Modifying Natural Systems with Extensive Use of Engineering Measures

Attempts to control debris flows and floods by modifying natural systems with engineer measures has long been the major component for the so called "watershed management " approach in Taiwan. Engineering measures, including reservoirs, levees, and channelization are used to modify the natural system for flood control, in contrast to flood proofing and floodplain management measures for altering human behavior to minimize impacts(Table 4). Engineering measures such as deposition areas, check dams, deflection walls (berms), terminal walls, berms or barriers, debris racks, grizzlies or other debris-straining structures, and storage basins are extensively used to cope with debris flow problems in Taiwan (Table 4). In general engineering structures should be used only in areas identified as having a high risk of debris flow impact. In August 2000, the government in Taiwan had decided to undertake a program worth NT$10 billion over three years to control debris flows in Taiwan. The main goal is to alleviate problems of the disastrous debris flows that will likely occur at sites with significant amount of materials already deposited in lower slopes and stream channels due to many landslides caused by the 921 earthquakes.

There is concern by some professionals and environmentalists that there may have been an over-reliance on engineering solutions to debris flow and flooding problems. Furthermore, some engineering measures can cause unwanted effects, particularly when considering cumulative watershed effects. The loss of hydrologic function and environmental costs associated with the loss of natural rivers, floodplains, and estuaries may also exceed the benefits for which they were constructed. For example, Leopold (1994) points out that intensive construction of levees since 1927 on the Mississippi river have increased flood stages – for the 1993 flood, the increase was 10 feet (3m) above what it would have been without levees at St. Louis, Missouri,
U.S.A. The emphasis on structural solutions to protect people from landslides, debris flows and floods has sometimes resulted in a false sense of security in people living downstream. Engineering structures may also be detrimental to stream ecosystems and habitats.

The trememdous damages and loss of lives caused by most recent floods and debris flows over many parts of Taiwan in July and September this year (2001) also indicated that most engineering measures had not been successful in achieving intended disaster mitigation goals under extreme hydrologic conditions (Table 4). This experience and lesson strongly suggest that the floods and debris flow mitigation programs that rely heavily on engineering methods should be more critically reviewed in terms of mitigating disasters and achieving sustainable land and water management. Engineering solutions to mitigations of disasters caused by hydrologic and tectonic extremes need to be carefully examined to ensure that (1) the intended effects are realized, (2) they are economically justified, (3) environmental impacts are minimized, and (4) people occupying hazardous areas do not develop a false sense of full security from structures that can provide only limited protection (Brooks, 1998).

Table 4. Examples of measures commonly used to mitigate problems associated with floods and debris flows (modified from Brooks et al. 1997, Fiebiger 1997, Lee et al. 1990).

Problems	Structural Measures	Non-Structural Measures
Floods	*Dams/reservoirs *Levees,Dikes *Channelization *Channel diversions	*Flood Proofing *Floodplain management *Maintain watershed vegetative cover; watershed management *Riparian management *Flood warning-forecasting
Landslides	*Retaining walls *Pilings *Buttresses	*Zone hazardous areas *Maintain vegetative cover with appropriate species; forest watershed management
Debris Flows	*Debris racks *Deposition areas *Deflection walls *Terminal walls *Berms and barriers *Storage basins *Debris flow traps *Check dams	*Hazard management & risk mapping *Forest watershed management *Warning system

b. Modifying Human Systems and Behaviors

It should be pointed out that floods and debris flows are natural phenomena. They become disasters only when people occupy their downward travel routes. Unfortunately, human encroachment on flood plains & hazardous areas with both controlled and uncontrolled development is a common phenomenon occurring in Taiwan. People on the island have quite often lost touch with their natural environment and are unaware of the danger and hazards of the area in which they live and work. Therefore, one thing that is fundamentally important is a change in man's basic attitude toward the land, hillslopes and stream channels. The basic principle is the we should not compete with debris flows and floods for spaces that are supposed to be their initiation zones, travel routes or deposition areas.

The first step is to determine where events such as landslides, debris flows, and floods are likely to occur, and then either restrict human activity in the areas deemed hazardous, or allow occupation, but provide an early warning system to allow people to escape disasters. Floodplain management represents the former approach in which degrees of risk are delineated on the floodplain on the basis of 25-, 50-, 100-yr or 200 yr floods. Human activity is then restricted according to the degree of risk. Flood warning systems are an example of the latter method, in which people then evacuate hazardous areas when weather conditions warrant such actions (Brooks, 1998). In addition to promoting more flood-plain management programs, many professionals are also calling for greater emphasis on protecting and restoring rivers and floodplains to their natural condition to complement structural and nonstructural measures already in place (Faber and Hunt 1994, Leopold 1994, Shabman 1994).

White it is highly desirable to first properly identify flood and debris flow hazard zones and then avoid human activity in the area completely, many hazardous areas in Taiwan have already considerable economic activity and extensive development occurring within them, particularly in areas with dense human populations. Moving human activity out of debris flows and floods hazardous areas in Taiwan may be very difficult due to economic and political considerations. For enhancing the usefulness and effectiveness of changing human activities (behavior) requires that appropriate policies and institutional arrangements be in place to foster closer cooperation between people and government institutions.

c. Coping with Uncertainties of Extreme Natural Events

In taking a flood plain management or zoning approach, we have to define the risk that certain extreme, natural events will occur. A long-term perspective that recognizes the uncertainty of events occurring within given time horizons is essential. This requires that we have excellent and long-term data sets that allow us to empirically relate past events with the probability of future events occurring. However, there is always a degree of uncertainty of prediction. The wide ranges of confidence limits around most calculated 100-500 year floods based on rather short period of data indicate high degree of uncertainty. Recent post-flood studies in Taiwan have raised questions about the adequacy of flood-frequency estimates. Reasons for discrepancies between estimated and actual frequency of occurrence can be attributed to one or more of the following: (1) insufficient number of years of data to define the true frequency curve, (2) an inadequate understanding of the effects of normal climatic variation and periodicity, and (3) long-term climatic change (Brooks, 1998). Examining historical hydro-meteorological records of Taiwan it was found that (1) the magnitudes of debris flows and annual floods tends to be increasing and (2) that the more rare floods, with recurrence intervals over 50 years, seem to occur several times over past 20 years. Rapidly changing climatic conditions possibly due to global warming and El Nino phenomena in recent years can add additional uncertainty to determination of peak flows. Furthermore we still do not know when such events will occur next, only the probability for a particular time period.

d. Disaster Mitigation Projects for Job Creation in Rural Area

After the September 21 1999 earthquake, many people in Taiwan, particularly aborigines living in mountainous areas are badly in need of jobs to earn income for supporting their families. Therefore, it is highly desirable to integrate disaster mitigation projects with other government programs in providing services to the people in need of assistance. A newly implemented initiative by the Council of Agriculture is to integrate properly designed job creation projects and soil conservation projects by recruiting volunteers and hiring local residents and aboriginal people. This initiative help achieve the broader goals of social stability, prosperity and environmental conservation and help participants establish a sense of satisfaction and confidence as they contribute to a worthy cause of environmental conservation at the local level.

e. Public Information, Education and Participation is Essential

People are becoming more aware of environmental matters and concerned about the condition of the world for future generations. This concern translates into increased political awareness, pressure, and action. It helps create more effective and acceptable means of distributing the costs and benefits of watershed management programs. To expand this awareness, public policy support is needed to improve and extend public education and training programs.

A properly designed and implemented public education and information dissemination sub-program is essential to help generate and sustain the understanding, acceptance and involvement of people living in areas most likely to be detrimentally affected by debris flows and to ensure successful implementation of mitigation measures. For example, after cautious consideration, relocation of the village debris flow hazard areas to safer locations may be the best option to avoid the disastrous impact of debris flows to lives and properties. However, it will take a significant amount of persuasion and effort, including incentives and compensation measures, before the villagers agree to the move of their villages to nearby safer locations.

Good public information dissemination and education efforts will encourage closer cooperation between people and institutions for achieving disaster mitigation goals. For example, it will greatly help implement the evacuation work and relocation plan for people living in debris flow and flood hazard zones. In addition, securing strong public support through public information and education is politically critical to ensure that disaster mitigation program is included in government's priority list, among many other essential government programs for funding and manpower.

Discussion

In Taiwan intense and long duration typhoon rainfalls, triggers landslides, debris flows and floods that cause loss of life and property damage annually. The impacts of these events are as much caused by human behavior on the watershed as the magnitude of the events themselves. Scarcity of land resources can dictate where people live on the landscape and what they do, in many cases making them more vulnerable to hydrologic and tectonic extremes noted above and further complicating the development of sustainable measures to cope with extreme events.

For reducing debris flow hazards, Davies (1997) suggested three options: (1) modifying the natural system, (2) modifying the human system of behavior, and (3) some combination of (1) and (2). The same options exists for mitigating landslides or flood hazards. In Taiwan it is essential to modify both the natural system and human behaviors. The following points need to be discussed for developing sustainable and effective disaster mitigation strategies in Taiwan.

a. Landuse and Disasters

Debris flow and flood hazards are the results of natural processes interacting with human systems. Of concern is the extent to which human "system" and specifically land use disturbances on the watershed contribute to such processes. In Taiwan, the cultivation of tea and betel nut palm on slopelands brings attractive financial returns to farmers. Although extensive areas are not cultivated throughout the island, the combination of steep slopes, intensive rainfall and erosive soils promotes serious, localized problems of landslides and high rates of surface soil erosion. Degraded lands also accelerate soil erosion and sedimentation rates, reducing the capacity of stream channels to transmit water flow during floods. When forest vegetative cover is removed from uplands for agricultural cultivation, soil erosion can exceed 220 t/ha/yr (Wu, 1989). Perhaps more serious is the loss of forest cover where soils are shallow, prompting landslides and debris flows during torrential rainstorms (Brooks,1998).

Cheng et al. (1997) suggest that improper slopeland cultivation, removal of vegetative cover, and road construction contributed to the landslides and debris flows. They point out, however, that such events are likely to occur any time rainstorms are extremely intense, or when they last 12 or more hours and total at least 150 mm, no matter what the land use. This is particularly true in many of the high mountain areas, where hillslopes often exceed 70%, channel gradients exceed 40%, and channels are loaded with material from landslides along the channel side slopes. Although, to some extent human activity on the watershed can contribute to the magnitude of such events, it is unrealistic to suggest that halting all human activity on such watersheds will prevent them from occurring. On the other hand, developing effective measures and responses in Taiwan is compounded by difficulties in implementing proper land use planning and regulation due to land scarcity, inadequate policies and lack of cooperation between people and institutions.

b. Limited Effectiveness of Disaster Mitigation Measures

Developing sustainable land and water management requires that we effectively cope with the risks and uncertainties of extreme events such as floods, landslides, debris flows, and droughts. To a limited extent, natural watershed systems can be modified through soil and water conservation and other measures to reduce the severity of such extreme events. However, *no matter how advanced our technologies*, there will still be natural occurrences of typhoons, floods, landslides, debris flows and droughts, and the degree to which such events result in disasters is in effect, due to human behavior and disturbance on the watershed. A set of management guidelines, contingency plans, and so forth, need to be in place that recognize the risks and uncertainty of floods and debris flows so that human populations are better prepared when they do occur.

People often have exaggerated expectations of the effectiveness of protective measures against floods and debris flows. For example, protection against floods by upstream reservoirs, levees, and channelization works imparts a false sense of security to those living in the floodplain. However, the effectiveness of there flood protection measures is limited and people living in flood plain must recognize that they are vulnerable to extreme events. We may somewhat reduce the degree of their impact, but we cannot completely prevent large floods and debris flows from happening. This concept of limited effectiveness may be difficult for politicians or some people in administrative authority to accept. However, it is important that it be recognized.

c. Watersheds as Disaster Mitigation Planning and Management Units

Watershed management is a viable concept and a practical means of working towards sustainable use of land and water resources. By using watersheds as units for analysis, many of the important ecological relationships and linkages with human activities can be more readily evaluated (Montgomery et al., 1995), thereby reducing the prospects of causing negative impacts. However, many questions remain in terms of quantitatively valuing and accounting for such effects, particularly when contrasted to the human and financial benefits and costs associated with disaster prevention objectives of watershed management (Brook, 1998).

Watersheds are ideal systems to understand the cumulative effects of human activities on land, water and related environmental entities for improved land and water management and reduced vulnerability of people to extreme weather and tectonic events. The effects of any one type of land use should be considered in the context of other uses of resources within the same watershed as they affect the flow of water and water-born sediment, and water quality. What seemingly minor incremental impacts individually can collectively become more serious when considered over time and space. These are referred to as cumulative effects of land use and management (Reid 1993). The changes in vegetative cover, soils, and hydrology of a watershed can have subtle changes from any one year to the next, but can become amplified over time. The position on the watershed in which certain changes occur, likewise can have greater or lesser effects (Brooks, 1998).

The upstream-downstream relationships are important in understanding impacts and in developing appropriate corrective actions. Floods with large amount of huge boulders that caused 9 deaths due to broken dikes in Chu-shan, Nantou County during typhoon Toraji on July 30, 2001, mainly originated from upstream tributary watersheds. After the major disasters caused by typhoon Toraji, the central government in Taiwan had recently decided to develop integrated watershed conservation and management plan for 4 major river basins in central Taiwan. This is an important step in the right direction. For long term perspective, it is also highly desirable to establish cooperative or joint political units in watersheds such as river basin commissions and watershed districts. This is particularly useful where different agencies have different responsibilities for land and water use on watersheds (Brooks, 1998).

There is an increasing awareness and understanding of on the hydrologic, geomorphic and ecological roles of natural streams, rivers, wetlands, riparian systems and floodplains as alternatives to structural (engineering) measures that have been so widely used in soil and water conservation projects in the context of extreme hydro-meteorological and tectonic event. Ecological objectives, including biodiversity values, are becoming more mainstream in planning and management. In many cases ecosystem values have become a target of management, rather than being viewed as constraints to management

d. Policy Considerations and Instruments

Using a watershed management framework to reduce people's vulnerability to debris flow and flood hazards through sustainable use of land and water resources makes sense from a biophysical perspective. However, supportive policies and institutional arrangements are required for encouraging people to convert these concepts into practice (Brooks et al. 1994).

County and township boundaries in Taiwan rarely coincide with natural watershed boundaries. The general public and politicians and development professionals generally are not fully aware of watershed relationships and the importance of upstream-downstream linkage in mitigation of debris flows and floods disasters. This further limits the development and enforcement of policies that promote disaster mitigation through sound watershed management.

A number of policy instruments are needed by all levels of government for encouraging improved conservation and management of land and water resources. Policy instruments can be grouped into three general categories: (1) regulatory, such as zoning, regulations, land and water rights, controls, permits, prohibitions, and licenses, (2) fiscal, including prices, taxes, subsidies, fines, and grants, and (3) direct public investment and management, such as technical assistance, research, education, land management, installation of structures, and infrastructure. Zoning areas within watersheds that are considered susceptible to hazards (floods, landslides, debris flows) may be used in conjunction with fiscal instruments, such as adjusting insurance rates according to degree of hazard. In essence, a penalty should be paid by those who carry out economic activity in vulnerable areas. In situations where land use upstream adversely affects downstream entities, policies are needed to make groups more responsible for their impacts on other groups (Brooks et al. 1994).

Conclusions

Structural (engineering), bioengineering and vegetative measures can be developed to exert some measure of control over floods and debris flows of small to medium magnitudes. However it is unrealistic to believe we can control large and extreme events. The possibility of imparting a false sense of security related to floods and debris flows should be clearly pointed out to politicians and general pubic alike.

For mitigation of floods and debris flows in land scarce, densely populated areas such as Taiwan, some combination of structural and nonstructural measures is needed. Some structural measures may continue to be components of watershed management, but more comprehensive criteria are needed to determine where such measures are and where they are not appropriate. To mitigate natural disasters caused by extreme hydro-meteorological and tectonic events, greater emphasis should be placed on modifying human behavior until we have a more complete understanding of natural systems and the extent to which we can realistically control the behavior of natural systems to extreme events.

Comprehensive assessments of the effectiveness of debris flow and flood mitigation options and of their economic benefits and costs are needed, but should be evaluated within a watershed perspective. More comprehensive criteria need to be integrated into planning and management that considers both the hydrologic function and multiple ecological benefits of natural rivers, floodplains, riparian systems and wetlands as complementary to engineering structures in developing disaster mitigation efforts. Appropriate policies and favorable institutional mechanisms related to sociceconomic, cultural and political factors—are required for effective implementation of disaster mitigation program.

References

Brooks, K. N., P. F. Ffolliott, H. M. Gregersen, and L.F. DeBano (1997) *Hydrology and the Management of Watersheds.* Second Edition. Iowa State Press, Ames, Iowa, 502 pp.

Brooks, K. N., P. F. Ffolliott, H. M. Gregersen, and W. Easter (1994) Watershed management and upland conservation: policy framework for sustainable development. Policy Brief No. 6. EPAT/MUCIA/USAID.

Brooks, K. N., H. M. Gregersen, A. L. Lundgren, R. M. Quinn, and D. W. Rose (1990) *Watershed Management Project Planning, Monitoring, and Evaluation: A Manual for the ASEAN Region.* ASEAN-US Watershed Project, Philippines, 215 pp.

Brooks, K, N.(1998) Sustainable watershed management in areas vulnerable to natural disasters: Challenges for the 21st Century. Proceeding of International Seminar on Watershed Conservation and Management April 22-23, 1998, Taichung. Taiwan. Water Resources Bureau, Ministry of Economic Affairs and National Ching Hsing University.

Cheng, J. D., H. L. Wu and L. J. Chen (1997) A comprehensive debris flow hazard mitigation program in Taiwan. Pp. 93-102 in: *Debris-Flow Hazards Mitigation: Mechanics, Prediction, and Assessment.* Proc. First International Conference, C. Chen (ed.). American Society of Civil Engineers, New York, NY.

Davies, T. R. H. (1997) Using hydroscience and hydrotechnical engineering to reduce debris flow hazards. Pp. 787-810 in: *Debris-Flow Hazards Mitigation: Mechanics, Prediction, and Assessment.* Proc. First International Conference, C. Chen (ed.). American Society of Civil Engineers, New York, NY.

Faber, S. and C. Hunt (1994) River management post-1993: the choice is ours. *Water Resources Update*. Issue No.94: 21-25. The Universities Council on Water Resources.

Falkenmark, M. (1997) Society's interaction with the water cycle: a conceptual framework for a move holistic approach. *Hydrological Sciences* 42(4): 451-465.

Fiebiger, G. (1997) Structures of debris flow countermeasures. Pp. 596-605 in: *Debris-Flow Hazards Mitigation: Mechanics, Prediction, and Assessment.* Proc. First International Conference, C. Chen (ed.). American Society of Civil Engineers, New York, NY.

Gregersen, H., K. Brooks, J. Dixon, and C. Hamilton (1987) *Guidelines for Economic Appraisal of Watershed Management Projects.* FAO Conservation Guide 16. Rome: Food and Agriculture Organization of the United Nations. 130pp.

Knox, J. C. and Z. W. Kundzewicz (1997) Extreme hydrological events, paleo-information and climatic change. *Hydrological Sciences Journal* 42(5):765-779.

Kundzewicz, Z. W. (1997) Water resources for sustainable development. *Hydrological Sciences - Journal-des Sciences Hydrologiques* 42(4): 467-480.

Lee, S. W. (1981) Landslides in Taiwan. Pp. 195-206 in: *Problems of Soil Erosion and Sedimentation.* Proc. South-East Asian Regional Symposium, 27-29 January, 1981. Bangkok, Thailand.

Lee, S. W., J. D. Cheng, C. W. Ho, Y. C. Chiang, H. H. Chen and K. J. Tsai (1990) An assessment of landslide problems and watershed management in Taiwan. Pp. 238-246 in: *Research Needs and Applications to Reduce Erosion and Sedimentation in Tropical Steeplands*. Proc. Symp. IAHS-AISH Publ. No. 192.

Leopold. L. (1994) Flood hydrology and the floodplain. *Water Resources Update*. Issue No.94:11-14. The Universities Council on Water Resources.

Meigh, J. R., F. A. K. Farquharson and J. V. Sutcliffe (1997) A worldwide comparison of regional flood estimation methods and climate. Hydrological Sciences Journal 42(2): 225-244.

Mimikou, M. A. (1995) Climatic change. Pp. 69-73 in: *Environmental Hydrology*. V. P. Singh (ed.). Kluwer Academic Publishers, The Netherlands.

Montgomery, D. R., G. E. Grant and K. Sullivan (1995) Watershed analysis as a framework for implementing ecosystems management. *Water Resources Bulletin* 31: 369-386.

Piechota, T. C., J. A. Dracup and R. G. Fovell (1997) Western US streamflow and atmospheric circulation patterns during El Nino-Southern Oscillation. *Journal of Hydrology* 201:249-271.

Reid, L. M. (1993) Research and cumulative watershed effects. USDA Forest Serv., Gen. Tech. Rep. PSW-GTR-141.

Rosegrant, M. W. (1997) *Water resources in the Twenty-First Century: challenges and implications for action.* Food, Agriculture, and the Environment Discussion Paper 20, International Food Policy Research Institute, Washington, D.C.

Rosegrant, M. W. and R. S. Meinzen-Dick (1996) Water resources in the Asia-Pacific region: managing scarcity. *Asian-Pacific Economic Literature* 10(2): 32-53.

Scherr, S. J. and S. Yadav (1996) *Land degradation in the developing world: implications for food, agriculture, and the environment in 2020.* Food, Agriculture, and the Environment Discussion Paper 14, International Food Policy Research Institute, Washington, D.C.

Shabman, L. (1994) Responding to the 1993 flood: the restoration option. *Water Resources Update*. Issue No.94:26-30. The Universities Council on Water Resources.

Sun, H. and D. J. Furbish (1997) Annual precipitation and river discharges in Florida in response to El Nino and La Nina-sea surface temperature anomalies. *Journal of Hydrology* 199:74-87.

Tobin, G. A. and B. E. Montz (1994) *The great Midwestern floods of 1993.* Saunders College Publ., Harcourt Brace College Publ., New York.

Wu, Huei-Long (1989) *Soil conservation on slopelands in Taiwan.* The Chinese Soil and Water Conservation Society. No. 8-1. Taipei, Taiwan.

Erosion Control and the Impact of Highway Construction on Wetland Water Quality: A Case Study

Chin-Pao Huang[1], M. ASCE, and Robert S. Ehrlich[2]

Abstract

To assess the efficacy of control on sedimentation and erosion during highway construction, the impact on water quality in adjacent wetlands was monitored. Water quality measures used were turbidity, suspended solids, conductivity, alkalinity, hardness, pH, organic carbon, dissolved oxygen, nitrate and phosphate. Heavy metal concentrations in surface waters and stream sediments were measured. Upstream measurements, used as controls, showed periodic, seasonal variations. Downstream measurements were not significantly different from upstream measurements except during an incident when erosion controls were neglected and during covert construction. Culvert construction resulted in resulted in increased sediment and turbidity from diversion and dewatering and increase in hardness, attributed to concrete leaching. Attention to sedimentation and erosion controls and seasonal scheduling of highway construction are advocated as a means of ameliorating the effects on wetlands.

Keywords: wetland, water quality, highway construction, erosion control

Introduction

The impact of highways upon the environment, in particular wetlands, has been a concern in the design and construction of highways (Bitter and Bowers, 1993). The location of the highway may require filling of wetlands, but the impact may be lessened by choosing a suitable route. A second concern is ameliorating the impact of highway runoff exacerbated by vehicular traffic (Smith and Kaster, 1983; Yousef et al., 1987). Researchers have studied the water quality aspects of highway runoff (Thomson et al., 1994) and the long-term ecological effects in a number of different environments (Ogbeibu and Victor, 1989; Childers and Gosselink, 1990). Others have evaluated the effects of short-term disturbances (Detenbeck et al., 1993; Tikkanen et al., 1994). Few studies have been done however, on the effects of the intermediate term highway

[1] Distinguished Professor, Department of Civil and Environmental Engineering, University of Delaware, Newark, DE 19717, USA.
[2] Graduate Research Assistant, Department of Civil and Environmental Engineering, University of Delaware, Newark, DE 19716, USA.

construction period itself. Darnell examined the qualitative impact of construction (Darnell, 1976). Water quality impacts of logging coupled with road construction have also been studied (Burns, 1972; Harr and Fedricksen, 1988; Grayson et al., 1993). Protocols for construction generally include specific instructions for sedimentation and erosion control (Delaware Department of Natural Resources and Environmental Control, 1989; Maryland Department of Transportation, 1989). These include silt fences between construction and streams, sedimentation basins in ditches leading into wetland streams and berrms to prevent loose material from washing from roadway surfaces into streams. The Delaware Department of Transportation (DelDOT) initiated this quantitative study to examine water quality impacts of highway construction and to evaluate the effectiveness of the required control measures. Measurements were made for a period of more than two and a half years.

The study was conducted at the expansion of an existing highway, U.S. Route 113. This primary four-lane divided highway in Sussex County, Delaware, has a two-lane section, which had become congested as annual average vehicular traffic exceeded 10,000 in 1989 (DelDOT, 1991). To relieve traffic congestion, improve safety, and to provide a more efficient and safer highway facility for future traffic along Rout 113, DelDOT was expanding this remaining two-lane section between Miford and Georgetown to a four-lane divided highway.

This construction required filling a total of 18 hectares of wetland within the project right-of-way which mitigating the effects of constructing wetlands at a nearby site. Following DelDOT guidelines for erosion and sediment control, measures were included in the construction contract to minimize the impacts at the construction site itself to those necessary for construction. The construction included pipe crossings and culverts, which prevent the road from acting as a dam that would damage the wetlands by alerting water levels (Stoeckel, 1967; Jeglum, 1975). Concern had arisen about the possible changes in hydrology of the Hudson Pond system as a result of highway expansion and their impact on the flora of the region. Therefore, we also included measurements of water levels in this drainage system.

Our general approach has been to make field measurements and observations at a number of sites. These were chosen to coincide with pipe or culvert crossings. We made additional water quality measurements on samples brought back to the laboratory. The study period began after the initiation of construction and hence, we evaluated the impact by comparing water quality a sampling points upstream and downstream of the highway right-of-way. We used field observations and construction schedule to determine the presence of different highway construction operations such as grubbing, fill, paving, culvert construction and removal of old pavement. These operations were then examined for correlations with changes in water quality.

Selection of Sampling Points

The project area is relatively flat, with significant changes in elevation are only where the highway crosses the water courses. Soils in the area are generally hydric with muck areas at Hudson Pond (DelDOT, 1991). The sampling points included in this study (Figure 1) were initially selected because they had water present through most of the study period. Other sites were initially selected but were not included because water was not present at both upstream and downstream sides of the highway during most of the study period. The selected sampling points and their characteristics are:

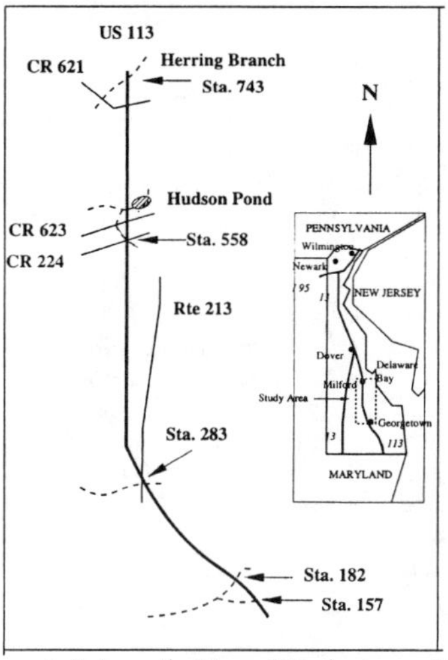

Figure 1. Schematic Map of Study Area. Streams are shown with dashed lines and roads with solid lines. The insert shows the study area in relationship to Delaware and surrounding states.

Station 157 is along Mifflin Ditch which drains westward to Deep Creek of the Nanticoke Basin and hence into the Chesapeake Bay. Forested wetlands in the vicinity are dominated by loblolly pine (*Pinus taeda*), red maple (*Acer rubrum*) and sweetgum (*Liquidambar styraciflua*). Soils are poorly drained hydric soils of the Fallsington and Pocomoke series. The existing 91-cm (36") reinforced concrete pipe was replaced by a 97x52 cm (38"x60") reinforced concrete pipe on a skewed alignment to the highway. The existing stream corridor was diverted, and a portion of it was filled to accommodate the alignment of the pipe. The total underwater fill area of this station is 237 m^2. The downstream bed was excavated to meet the elevation of the new pipe. Sampling points were outside of the right-of-way on the east (upstream) and west (downstream) sides of the highway.

Station 182 is also associated with Mifflin Ditch. Vegetation and soils are similar to those of station 157. The existing 3.6 m (12') concrete box culvert was removed and replaced on the same alignment by one of similar structure and slightly larger dimension. The total underwater fill area at this station was 116m^2. Excavation downstream (west) of construction occurred in 1993 to align the stream with new culvert.

Station 283 is located where Gravelly Branch crosses U.S. 113 near the intersection of U.S. 113 and CR 213. Gravelly Branch is associated with the Nanticoke River basin. Swamp waters from the Redden State Forest drain through this station. The wooded wetlands at this location include loblolly pine, red maple, sweetgum, possumhaw viburnum (*Viburnum numdum*), and high bush blueberry (*Vaccinum corymbosum*) (Delaware DelDOT, 1991). The soils are the Elkton series, and the seasonal high water table is 0-30 cm from the ground surface. The existing 6.1 m long x 1.2 m square (20'x 4') concrete box culvert was removed and replaced by one of similar size, structure, and alignment. The total underwater fill area at this station was 278 m^2. The upstream sampling site was on a path beside the stream, 50 m from the culvert. The downstream sampling site is off CR 213, about 20 m from the culvert.

Station 558 is associated with the floodplain of Schoolhouse Branch, which crosses from the east, then crosses CR 224 and flows north into the Hudson Pond system. This station is of particular interest since swamp pink (*Helonias bullata*), a threatened plant in Delaware, has been located about 50 m upstream on the east side of the roadway. This area is also a red maple swamp with sweet bay (*Magnolia virginiana*), American holly (*Llex opaca*) and greenbriar (*Smilax sp.*) abundant. During a preliminary assessment prior to construction by DelDOT there was no water in the stream channel. Fill in this area totaled 0.27 ha. Construction included regarding and paving the adjacent section of CR 224. The upstream sampling site was at the shoulder of the east side of the highway. Samples taken 100 m upstream in the wetland were similar to those at this point. The downstream site was adjacent to CR 224, about 50 m from the highway.

CR 623 is upstream of the highway on Schoolhouse Branch just above its confluence with other streams, which empty into Hudson Pond. We chose this site to get water, representative of the area upstream of Hudson Pond. Between the water crossing and Route 113 there was a single residence and ground was cleared for another one during the project period. This area includes loblolly pine, sweetbay, American holly and Atlantic white cedar (*Chamaecyparis thyoides*) and is close to agricultural properties and large, wooded residential plots. Construction included some regarding of this dirt road and addition of riprap at the metal pipe crossing.

Hudson Pond was formed by damming the stream below the roadway. Standing water exists year round and soils are muck. Dominant plants are Atlantic white cedar, smooth alder (*Alnus maritime*), sweetbay and in the under story, sweet pepperbush (*Clethra alinfolia*) and cattails (*Typha latifolia*). Swamp pink has been found about 200 m upstream of the roadway on private property. Construction consisted of a 6.1 m long x 1.8 m square (20'x6') box culvert as an extension of the existing culvert. The culvert on the east side (north bound roadway) was originally considered sound and was to be kept. Cracks were later found in this culvert and it was reinforced but DelDOT had not yet decided to replace it during the study period. While there was concern that culvert construction might impact the swamp pink, which is dependent upon a narrow range of hydrologic conditions, and analysis by DelDOT indicated that the culvert would not change the upstream hydrology. Wetland fill was 0.31 ha plus about 158 m^2 of underwater fill.

CR621 is upstream Station 743 and about 500 m from the right of way. The stream at this site is bordered by red maple woodland. Residential and agricultural activities are located on either side and upstream of the sampling point. A salvage yard is located southwest of this point between CE 621 and Route 113.

Station 743 is on Herring Branch in the Mispillion River, which drains into Delaware Bay. The swamp area has Atlantic white cedar and smooth alder (*Alnus serrulata*) with red maple and dogwood (*Cornus sp.*) up slope. The soil is a Johnston silt loam with large amount of organic matter. The construction extended an existing 3 m x 1.2 m (10'x4') box culvert. Wetland fill was 0.2 ha.

Field Measurements of Water Quality

Field measurements were made of temperature, pH, conductivity, turbidity, and dissolved oxygen using a Solomat WP803 monitor with 803PS snode. The monitor was calibrated for pH in the laboratory with standard pH 7 and pH 4 buffers (Fisher Scientific). Turbidity calibration was performed with dilutions from a formazin solution with 1000 NTU. The platinum resistance thermometer was calibrated by the manufacturer. The conductivity calibration used freshly diluted water to set 0 point and a 0.005 M KCl solution to set the high point at 718 μS/cm. Earlier measurements were made in the laboratory with Markson conductivity meter. Dissolved oxygen calibration used aerated tap water as the saturated standard and a 6% sodium sulfide solution as the zero dissolved oxygen standard.

Sample Collection and Preparation

We collected surface water samples monthly from the U.S. Route 113 sampling stations. Grab samples were taken near midstream and at about mid-depth in 1-liter acid-washed Nalgene bottles. Samples from Hudson Pond were taken within 1 m of the pond edge. The bottles were stored on ice, and transferred to a refrigerator at 4 °C upon return to the laboratory. Pretreatment of the samples included filtration and acidification, and the treated samples were put in new acid-washed Nalgene bottles and stored in the refrigerator for subsequent analysis.

Sediment samples were taken from the top 7 cm of sediment of the streambeds using a corer constructed from a 100-mL syringe or directly into Nalgene boxes with tight fitting lids. Only the top 1 cm of sediment was analyzed following culvert completion. These samples were taken to assess the more recently deposited sediment.

Laboratory Methods

We used standard method for water quality analysis (Adams, 1990; APHA et al., 1985). Total alkalinity was determined by titrating 100-mL of the untreated sample with 0.2 M HCl. The sample was titrated to pH in the range 4.3 to 4.7. For samples with alkalinities less than 20 mg/L, the pH was lowered by another 0.3 pH unit.

Gravimetric analysis was used to determine suspended solids. Filtration was used to separate the suspended on a Millipore type HA 0.45 μm filter membrane.

Dissolved meals were determined after addition of 5 mL HNO_3 to 100 or 200 mL of water. The samples were heated and evaporated until the volume was about 10 mL. The samples were analyzed using flame aspiration with the Perkin Elmer Model 5000 atomic absorption spectrophotometer, and the metal concentrations were determined from separately constructed calibration curves using 3 to 4 samples prepared from slats of the metals.

Metals (Pb, Co, Ni, Fe, Cu, Zn, and Cd) in sediments were determined after a sequential extraction procedure (Lake, 1987; Tessler et al., 1979). The freely exchangeable metal was first extracted with 1 M ammonium acetate at pH 7. Additional metal was ten extracted with concentrated nitric acid. To 3 g of sediment it was added 3 mL of concentrated nitric acid. The tube was capped; but a small space was left to prevent build up of excess pressure. The tube was then boiled in a water bath for 1 hour. Hydrogen peroxide was then added in increments of 0.5 to 1mL. And boiling resumed between additions. Hydrogen peroxide addition continued until no gas was liberated upon addition. The maximum amount added was 3 mL. The samples were cooled and diluted to about 10 mL before analysis. A blank without sediment was treated in the same manner. "Total metal" was determined by atomic absorption of a nitric acid-peroxide extract of ash sediment. The nitric acid-peroxide treatment should remove all metal from organic and inorganic fractions except for metal bound as silicates, which would require HF treatment. This silicate fraction is not expected to be bioavailable in most soil systems.

The experimental procedure for total phosphate in the water phase has three steps: (1) sample pretreatment, (2) strong acid digestion, and (3) phosphorus determination. Samples for total phosphate measurement were not pretreated but measurements were made soon after collection. Thus, the reported measurements include phosphate bound to suspended solids. The second step involved the conversion of the acid-hydrolyzable forms of phosphorus to orthophosphate by the persulfate digestion method (Adams, 1990; APHA et al., 1985). Then total phosphorus, in orthophosphate form, was determined by the ascorbic acid colorimetric method (Adams, 1990; APHA et al., 1985). All glassware used in this experiment was soaked in acid and rinsed with distilled water. The absorbance of samples was red on a Hach spectrophotometer. At least some measurements in each series were done in triplicate on a single sample and duplicate preparations were also made. Variations between measurements were less than 0.03 mg/L.

Nitrogen was determined as nitrate and ammonia using Hach test kits (detection limits 0. mg/L and 0.015 mg/L respectively). Nitrate nitrogen was determined by cadmium reduction method. A nitrate standard was used to calibrate the instrument. Measurements on duplicate samples varied by les than 0.3 mg/L. Ammonia was determined using Nessler reagent and colorimetry.

Total organic carbon was determined with a Beckman analyzer. Potassium hydrogen phthalate was used to make standards in the range to 240 mg/L. Samples were prefiltered through a Millipore 0.45 mm filter before injection into the analyzer. A standard curve was determined for each set of measurements.

Biological Observations

On each sample collection day we observed the vicinity of the collection site over a 15-minute interval for the presence of birds, amphibian, fish and aquatic insects. We examined the streambed for the presence of emergent vegetation and algae. No quantitative measurements of numbers or species were made. During the period from March through October, at least ten aquatic insects were seen at all unaffected sites. Since the health of Atlantic white cedar is of prime regulatory concern, individual trees outside the construction zone were examined for signs of distress. Our observations were supplemented by interview with knowledgeable residents.

Control Measurements of Water Quality in the Absence of Construction

Background measurements of water quality indicators were for each of the sampling points. Since these measures may show day-to-day variations that are greater than the differences due to construction activities, we paired measurements at upstream stations with same day measurements at downstream stations. None of the reported measurements were made during heavy rainfall, which could cause large, but transient, changes in water quality. Figures 2 to 5 show that there were indeed fluctuations in these indictors at the upstream stations. These variations appeared to have a seasonal character. A sinusoidal curve with a period of one year provides an approximate indication of the seasonal variation ($r^2 \sim 0.6$). Seasonal variation was also assessed by ANOVA analysis, which indicated significant differences by the Scheffe criterion.

As seen in these figures the minima and maxima of each of the measures occurred at about the same period for all stations. Turbidity and suspended solids concentrations showed peaks in late summer. Analysis of the data showed no significant correlation wit rainfall occurring in the 2, 7 or 14 day periods preceding the measurements. Conductivity reaches maximum values in late winter or early spring and minimum values during summer or early fall.

This might be expected if the increased conductivity during early spring were due to leaching of ions from the soils by increased rainfall and groundwater melt in the upper soil layers. This is suggested for station 157 by a significant correlation between depth of surface waters and conductivity ($p < 0.001$). However, a smaller inverse correlation ($p = 0.08$) with water depth was found for station 283. Alkalinity peaked during the late summer/early fall at the same time as pH increases. Hardness showed no significant seasonal variation. These seasonal differences could reflect differences in temperature, water level, rainfall or decay of organic matter but no clear correlations were evident for the period of study.

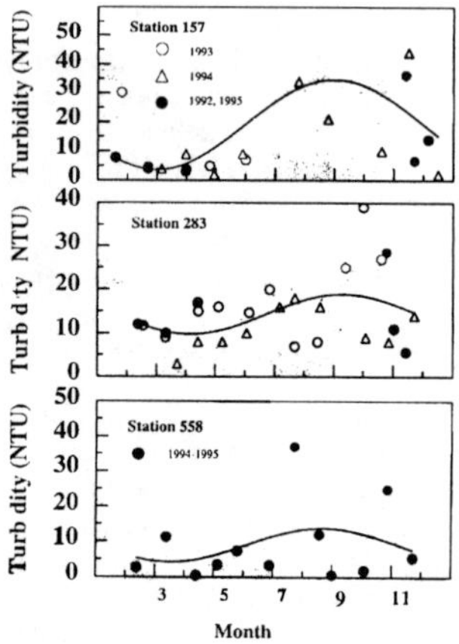

Figure 2. Variation of Turbidity by Month for Three Upstream Sampling Stations. The data for station 283 are delineated by year. The data for station 157and 182 are identified by year with symbols given in the graph of station 157 data. Data for station 558 were obtained between November 1992 and March

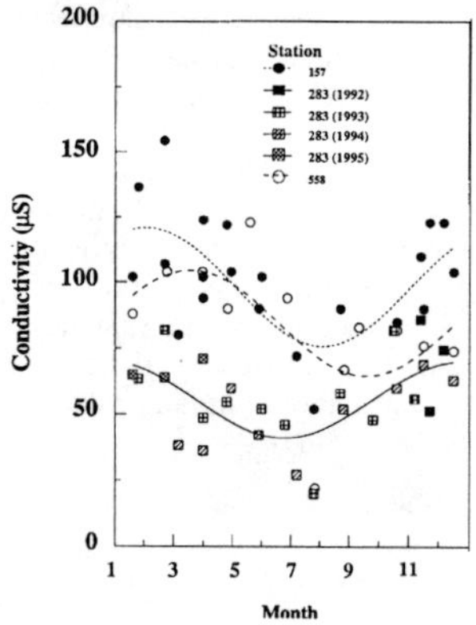

Figure 3. Variation of Conductivity by Month for Three Upstream Sampling Stations. The data fro station 283 are delineated by year. Data for station 157 were obtained between November 1992 and March 1995.

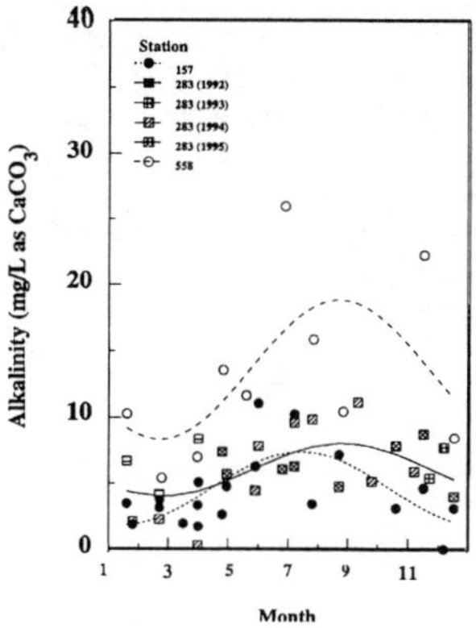

Figure 4. Variation of Alkalinity by Month for Three Sampling Stations.

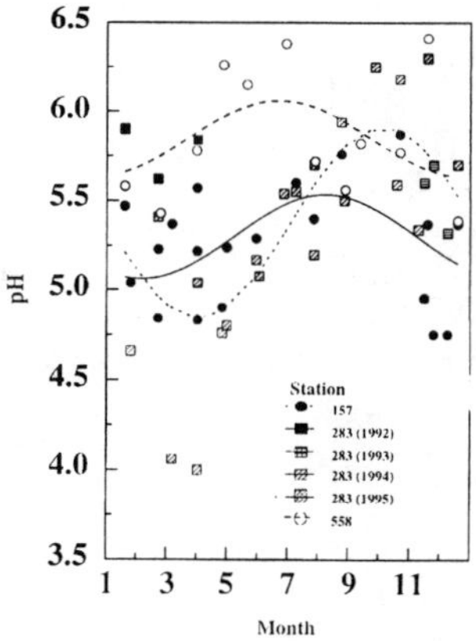

Figure 5. Variation of pH by Month for Three Upstream Sampling Stations.

Upstream-downstream Differences in water Quality in the Absence of Construction

The measurements of water quality parameters were separated into two groups based upon field observations of the presence of construction activities in the vicinity of the sampling stations. When available, construction schedules were also used to delineate period of construction. The habitat and soil may have differed across the highway for reasons other than recent anthropogenic activities. Differences may have arisen from past activities. In many places the wetland maps indicated that wetland areas lay parallel to the highway in lower areas arising from highway fill or followed ditches that were constructed to drain nearby farmland.

Tables 1 to 3 give the values for upstream and downstream measurements along with the results of statistical analysis. A paired t-test was used and hence only measurements were included for days when samples could be obtained from both sides of the construction area. In some cases one side was not accessible, was dry or was too shallow to obtain grab samples or use the water quality probe.

Table 1. Turbidity and Suspended Solids for Paired Sampling Points.

Sampling Point	Turbidity (NTU)			Total Suspended Solids (mg/L)		
	$x\pm\sigma^{\alpha}$	P^{b}	Range	$x\pm\sigma^{\alpha}$	P^{b}	Range
157E	15±19(17)	0.31	2-71	8±7	0.31	2-29
157W	18±22		2.7-79	10±14		1-56
182E	39±28(15)	0.42	1.1-105	14±21	0.42	1.3-86
182W	41±21		3.7-79	13±17		0.9-70
283E	38±34(17)	0.62	4-144	14±9	0.62	3-39
283W	37±26		10.2-112	13±7		5-30
558E	6.7±11.1(19)	0.87	0.5-24.8	46±11	0.23	27-59
558W	6.3±3.0		2.3-11.6	48±11		30-68
CR623	2.6±3.4(10)	0.16	0-11.5	47±4	0.59	42-55
Hudson Pond	5.1±6.2		0-19.2	45±5		36-54
CR621	8.6±12.2(10)	0.53	0-32	104±6	0.77	96-113
743	16.9±35.6		1.1-26	102±11		74-115

[a]: Mean ± Standard Deviation; [b] :Two-tailed paired t-test.

Table 2. Conductivity and Hardness at Paired Sampling Points.

Sampling Point	Conductivity (μS/cm)			Hardness (mg/L as $CaCO_3$)		
	$x \pm \sigma^{\alpha}$	P^b	Range	$x \pm \sigma^{\alpha}$	P^b	Range
157E	103±22(18)	0.002	56-155	47±27	0.31	2-29
157W	119±26		74-168	56±32		1-56
182E	47±13(15)	0.34	22-72	16±6	0.42	1.3-86
182W	53±17		23-78	17±7		0.9-70
283E	57±17(16)	0.99	20-86	14±9	0.62	3-39
283W	57±18		26-82	17±8		5-30
558E	64±25(12)	0.05	22-122	46±11	0.23	27-59
558W	98±38(18)		32-183	48±11		30-68
CR623	117±18(10)	0.41	107-165	47±4	0.59	42-55
Hudson Pond	101±37(12)		78-169	45±5		36-54
CR621	152±44(10)	0.93	58-217	104±6	0.77	96-113
743	147±51(11)		56-241	102±11		74-115

[a]: Mean ± Standard Deviation; [b]:Two-tailed paired t-test.

Table 3. Measured pH and Alkalinity for Paired Sampling Points.

Sampling Point	pH			Alkalinity (mg/L as $CaCO_3$)		
	$x \pm \sigma^{\alpha}$	P^b	Range	$x \pm \sigma^{\alpha}$	P^b	Range
157E	5.25±0.38(18)	0.18	4.75-5.88	4.75±5.88	0.03	0.1-11.1
157W	5.32±0.29		4.93-5.83	5.9±2.8		13-126
182E	4.63±0.96(17)	0.006	3.3-5.82	0.02±0.08	0.08	0-0.3
182W	4.81±1.06		3.34-6.07	2.9±5.3		0-17.4
283E	5.34±0.59(19)	0.41	4.0-6.3	6.2±2.6	0.06	0.4-11.2
283W	5.36±0.59		4.03-6.41	7.0±3.3		3.0-12.4
558E	5.85±0.36(12)	0.63	5.39-6.45	13.1±6.6	0.72	5.4-26
558W	5.87±0.21		5.39-6.26	13.8±9.4		5.6-41.3
CR623	6.39±0.33(10)	0.29	5.63-6.26	14.3±2.8	0.007	10.8-19.6
Hudson Pond	5.91±0.38(12)		5.43-6.56	17.0±4.0		11.4-24.9
CR621	6.39±0.33(10)	0.02	5.95-7.09	29.4±9.3	0.11	16.8-47
743	6.03±0.36		5.36-6.53	26.3±.6.5		19.8-37.3

[a]: Mean ± Standard Deviation; [b]:Two-tailed paired t-test.

In the absence of construction work there were no significant differences ($p > 0.05$ by a two-tailed paired t-test) in turbidity or suspended solids between any pairs of upstream downstream stations (Table 1). Significant differences at the 95% confidence level were seen at station 157 for conductivity and at station 182 (Table 2) and Herring Branch (CR 621 and station 743) for pH (Table 3). The conductivity, calcium and sodium at station 157 were higher at the downstream station but the cause was not determined. The hardness (Table 2) measured downstream at station 182 and 283 is higher than the upstream value by a small amount. This may have been a result of prolonged leaching from concrete following culvert construction at these sites.

The water at station 182 had little buffering power as indicted by low alkalinity and acidity (Table 3). The downstream channel was disturbed by excavation required to align the channel with culvert. As a result the fraction of organic matter in the sediment was decreased. Since decaying organic matter may lower the pH by production of organic acids, its paucity near the downstream collection point might allow the pH to increase. The decay of organic matter is an important factor in determining the pH of the surface waters of these wetlands and contributes tannins, which give the waters their reddish color. The decreased pH in Herring Branch downstream of the highway isles readily explained but may be influenced by the presence of a salvage yard south of the stream on the stream on the upstream side of the highway. The small changes in water quality indicators seen under control conditions, i.e., the absence of nearby construction, must be compared with any changes seen as a result of construction.

Water Quality Changes During Construction – Turbidity and Suspended Solids

Turbidity and suspended solids may affect wetland integrity by reducing the light available beneath the water surface. This would have a direct effect on submerged plants. Many researchers have reported the effects of suspended sediments on stream communities (Eaglin and Hubert, 1993; Tikkanen et al., 1994). Increased turbidity can result from increased stream velocities, which suspend clays from the streambed or introduction of fine sediments from the highway. Sedimentation of these suspended solids changes the nature of the streambed reducing seedling emergence (Jurik et al., 1994) and may deprive organisms of oxygen supply. High levels of suspended solids reduce the biomass (Smith and Kaster, 1983) and diversity (Ogbeibu and Victor, 1989) of benthic macro-invertebrates. Increase in suspended solids can also arise from changes in stream velocities that mobilize larger particles from the streambed as well as sediment washed into the stream from the highway or highway fill.

Figure 6 shows differences in turbidity between upstream and downstream values for turbidity at the sites studied. In the absence of local construction activity the average differences are close to zero ($p > 0.05$ by t-test of means). In order to more closely identify effects f construction, the differences in the presence of construction activity are divided into two categories. The first includes roadbed preparation: grubbing, excavation, grading and filling. The second involves operations for highway completion: those that involve concrete pouring or smoothing. Culvert construction, paving and pavement grinding have been included in this category. Figure 6 indicates that, in general, there are only small effects of roadbed preparation activities on stream turbidity. A notable exception is a single measurement at Herring Branch.

Field observation revealed that the increase in turbidity downstream of the highway was due to placement of fill without a sediment trap or even silt fencing between the roadway and stream. Sediment washed into the stream over a period of about a week. At the time of observation, late January, no aquatic insects were present in the stream before or during the increase in turbidity. During summer months we noted that another site, not part of the current study but with similar ecology, a turbidity increase of this magnitude resulted in the absence of aquatic insects found at an upstream location. In other operations some sedimentation and erosion control device was in place. These results indicate that, at least in the absence of storm events, the sedimentation control measures are effective in preventing turbidity increases in the wetland streams from roadbed preparation. At some locations silt fence was placed across drainage areas, leading to buildup of considerable sediment behind the fences. Under sheet flow conditions these fences may fail. While this was not observed during our study it should be noted that these fences are not substitutes for sedimentation traps or basins.

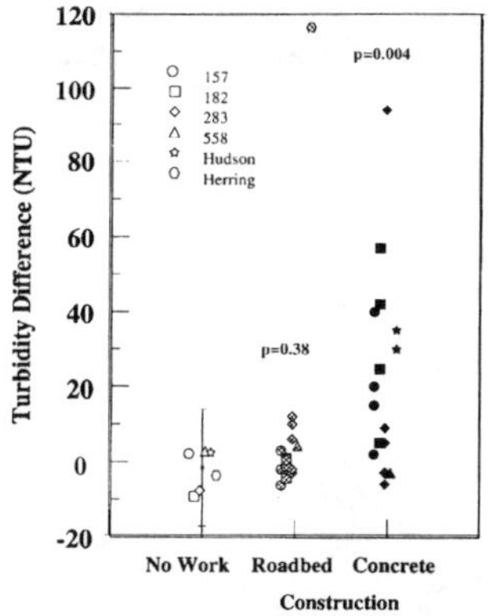

Figure 6. Differences in Turbidity between Downstream and Upstream Sampling Points. The points for periods when there was no construction work are average of 10 to 25 measurements. The errors bar shows the standard deviation for all measurements. Individual measurements are given by diagonally hatched symbols for work classified as Roadbed (clearing, grubbing, filling) and by solid symbols for Concrete (paving, pavement finishing, culvert construction). The probability indicted is from a t-test of the means. All differences are not significant ($p < 0.05$).

The operations designated "concrete" show significant increases in turbidity for many of the samples ($p < 0.05$ by a t-test of the mean). The larger differences occurred during culvert construction. Culvert construction requires diversion of the stream through a pipe, which generally increases the flow velocity. In addition the construction site is dewatered to allow pouring of concreter into the prepared forms. This dewatering also increases the flow at the outlet on the downstream side of the highway. This flow is passed over an outlet protection section of riprap to reduce its effects but in many cases the flow was diverted around the section or the riprap was inadequate to prevent scouring of downstream streambed. Also in some instances, the water from the dewatering operation was not screened to remove fine sediment and this contributed to downstream turbidity. The affected area was bout 20-50 m downstream of the right-of-way. Culvert construction activities lasted from a month to two months. The turbidity returned to normal a month after active construction.

The effects of construction on suspended solids concentrations are similar to those observed for turbidity (Figure 7). The deviation at Herring Branch due to the single unprotected operation is even greater as may be expected since the fill is sand and fine sand whereas the streambed sediment is a mixture of sand, clay and silt. No significant sedimentation was observed after two months, indicting that the sediment was dispersed over a wide downstream area.

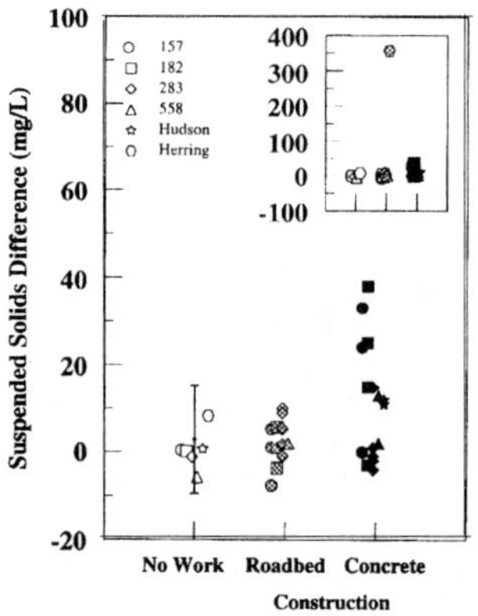

Figure 7. Differences in Suspended Solids Between Downstream and Upstream Sampling Points. The presentation is the same as Figure 5.

Water Quality Changes During Construction –Dissolved Substances

Dissolved substances were measured by conductivity, pH, hardness and specific measurements of heavy metals, oxygen, nitrate and phosphate. Conductivity (Figure 8) is a measure of the total dissolved ions and could change as a result of sediment induction, leaching or cement plant operation. Conductivity changes are not a problem except when they result from salt content beyond the tolerance of local floral and fauna. No significant conductivity changes were observed as a result of construction operations. Organisms are generally adequate to a specific range of pH and long-term changes in pH may favor new organisms including those not native to wetland habitats. Figure 9 shows the pH changes during construction. The effects of road bed operations are minimal ($p < 0.05$ using t-test of the means). Large changes were seen in pH during culvert construction particularly at station 182 which had low alkalinity and acidity. The changes brought the pH from very low values to values closer to neutrality. The altered values are close to the Delaware standard of 6.5-8. But the effects of such intermediate term changes on fauna adapted to low pH are unknown.

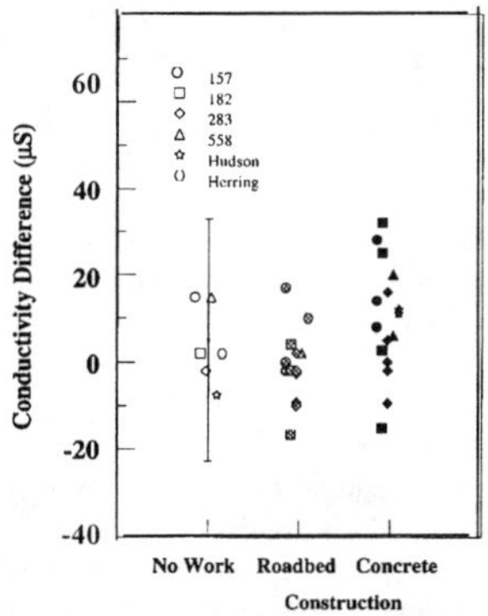

Figure 8. Differences in Conductivity Between Downstream and Upstream Sampling Points. The presentation is the same as Figure 5.

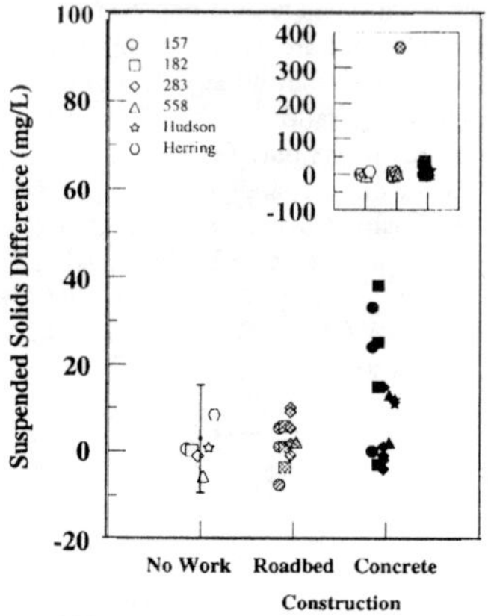

Figure 9. Differences in Ph Between Downstream and Upstream Sampling Points. The presentation is the same as for Figure 5.

Alkalinity and hardness depend on pH and reflect the presence of dissolved carbonate, calcium and magnesium. Alkalinity is a measure of the water's stability to respond to changes in acid content. High alkalinity is considered desirable and Delaware surface water standard specify a minimum of 20 mg/L as $CaCO_3$ (Delaware Department of Natural Resources and Environmental Control, 1993). Nevertheless, as the control values (Table 3) show, the pH and alkalinity of wetland waters is low due to the lack of dissolvable minerals and the presence of much decayed and decaying organic matter. Hardness (Figure 10) is a significant indictor of the bioavailability of heavy metals and metal criteria are a function of hardness (Federal Registrar. 995). The hardness differences between upstream and downstream stations are small. Significant deviations may be seen during paving operations (Figure 10). The increases in calcium and magnesium may be attributed to leaching of newly poured concrete. Comparison of downstream hardness values in the six months following any concrete operation with values at other times indicated a significant elevation ($p < 0.03$) while at the upstream points there was no difference ($p = 0.5$).

Nitrate and phosphate contents of wetland waters are significant since these substances are generally limiting nutrients. Excess amounts, particularly phosphate, can lead to algae growth and eventually to reduction in dissolved oxygen. Reduced oxygen in turn leads to destruction of invertebrates and fish kills. Measurements of dissolved oxygen were satisfactory (<50% saturation) throughout the measurement period. Nitrate and phosphate levels were elevated at some stations but, generally, no difference between upstream and downstream points were found. Elevated values probably result from agricultural operations or domestic septic systems. The only increase in nitrate was found at station 558W. This station is near a field office and the elected values could indicate leakage from septic devices there.

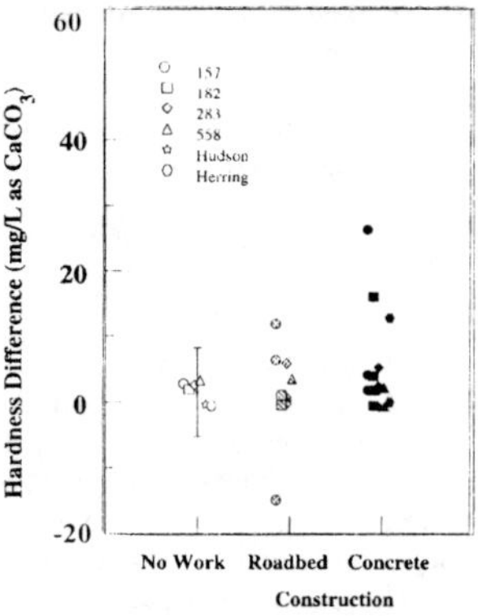

Figure 10. Differences in Hardness Between Downstream and Upstream Sampling Points. Presentation is the same as for Figure 5.

Metals in Water Sediments

A possible consequence of construction activity is the introduction of toxic heavy metals into wetland streams and sediments. Metal concentrations in surface waters were determined. All values were below the acute criteria set by the state of Delaware (Delaware Department of Natural Resources and Environmental Control, 1993). Within experimental error, no differences were found between upstream and downstream measurements.

The water values are instantaneous values and may not reflect influxes that occur over the construction period. Sediment samples were collected for analysis and the results are given in Table 4. At station 182, there is a significant difference ($p < 0.002$) in organic content between upstream (182E) and downstream (182W). This apparently arose from the channelizaiton downstream to align the stream with the new culvert. At station 283, after culvert construction, the top layer of sediment was removed by scouring. This layer reformed over a several month period due to transport from upstream. This top layer, a reddish, easily disturbed fraction, was analyzed separately and is the samples with high organic content. The total (acid extractable) concentration of zinc in the samples was found to vary linearly with organic content above a threshold of about 1% ($r^2 = 0.77$). Since the iron and organic contents were correlated ($r^2 = 0.66$), it cannot be ascertained whether zinc is bound to iron oxides, organic mater or both components. There is less correlation of other metals with either iron or organic content. For all sites the downstream metal levels were similar to those at upstream sites. Samples were also taken from the ditch, which collects runoff from the concentration site. The metal levels in the ditch are higher than those from the stream sediments with the exceptions of the surface fraction with very high organic and iron contents. This indicates that ditch sediments have the potential of increasing the heavy metal burden of the streams and it is prudent to take measures to prevent their intrusion.

Exchangeable metal levels cannot be easily correlated with the acid extractable levels but do show an increase inorganic content, downstream values are about the same or less than upstream values.

Hydrology

Hydrologic changes may result from highway construction when the highway acts as a dam between two sections of wetland. This is not a problem with the current project, however, since sufficient channels have been provided to carry water across the highway. A concern of the highway planners was that the extension of culverts to pass below two dual roadways would alter the upstream hydrology. This was considered undesirable because a threatened plant, swamp pine (*Helonias bullata*) was found upstream of Hudson Pond and upstream of Route 113 on Schoolhouse Branch (station 558). This plant was considered vulnerable to stress of water level changes.

Table 4. Metal and Organic Content of Sediments from Selected Stations.

Sampling Point	Organic Content (%)	Acid Extractable Metal (mg/kg)			Exchangeable Metal (mg/kg)
		Cu	Pb	Zn	Zn
182E	2.77±0.82	1.0±0.12	5.9±0.9	3.1±1.0	94±36
182W	0.58±0.26	0.98±0.61	3.1±3.1	4.2±3.4	113±57
283E	2.08±0.21	2.0	4.0	10.4	400
	29	2.6	2.5	94	600
283W	0.13	1.1	2.1	3.7	130
	1.71	0.7	0.9	8.5	150
	8.72	3.7	7.0	24	720
	18.2	72.3	16.5	72.3	--
558E	1.4	0.53	9.7	6.4	120
CR623	2.0	1.0	1.3	14	250
	8.7	1.0	1.3	14	250
	15.1	7.5	22.5	22.1	2950
743	25.8	4.0	3.9	43	92
Ditch	4.4	4.5	11.2	11.1	190

Water level measurements were made when collecting samples at two points in the Hudson Pond basin, station 558W and CR 623. In addition the changes in hydrology upstream of station 743 was measured at CR 621. Two measurements were made at each point. The depth of the water was measured and the level of water measured from the fixed point of the top of the pipes carrying water under the roadway. Data are shown in Figure 11. The data show that the water level upstream of Hudson Pond was the highest at the time of the first measurement. The water level at 558W was at the top of the pipe. Changes due to general regional variation can be estimated from the changes at CR 621 and from flow data at nearby Beaverdam Branch in the Mispillion Basin (Us Geological Survey, 1994). Unfortunately, no pother measurements were made prior to commencement of work on the Hudson Pond culvert. The water level dropped and remained almost constant after the work on the culvert was completed. Flow data on Beaverdam Branch (and also in the Nanticoke system) indicate that the period of measurement does not coincide with any period when flows in this area are greater than normal. Thus the temporary increase in water level at Schoolhouse Branch is probably attributable to the downstream construction.

Water depth also changed little except at 558W where sediment was found to have washed into the stream. The likely source was fill piled in a berm along CR 224 with no silt fence between the roadway and the stream. Erosion of this fill was not actually observed but erosion of similar berms was noted elsewhere.

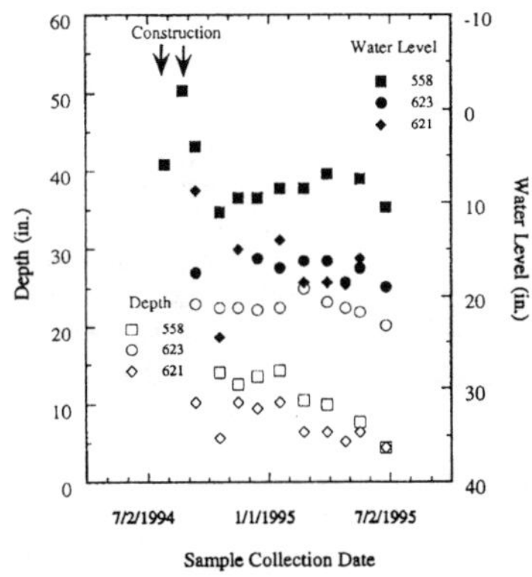

Figure 11. Variation in Water Elevation and Depth for Two Sampling Points (558W and CR 623) in the Hudson Pond Watershed and One Along Herring Branch (CR 621).

Field observations indicate limited biological consequences of construction activity. The threatened plant, swamp pink, was still observed in the region upstream of Hudson Pond after the culvert construction. No damage to Atlantic while cedar was observed. A systematic survey of fauna prior to construction was not available and a systematic survey was not made in his study. Residents of the region report a decrease in wildlife that use both wetland and upland areas as a result of the construction activities and operation of a concrete plant, immediately north of the construction area. Turkey, deer, ospreys and otters have disappeared or decreased in number. Hudson Pond is home to mallards and a single "tame" Canada goose. These have remained following construction.

A necessary phase of culvert construction is the use of sheet piling to channel water around the construction area. At Hudson Pond this piling has been left in place after completion of the southbound roadway with a small weir for water flow. In all of the most recent observation visible to the site, we observed turtles and fish attempting to swim upstream but impeded by the sheet piling. This may prevent their access to upstream food sources and breeding areas.

Conclusions

The water quality in the study region did not decrease below State of Delaware Standards during roadbed preparation except when there were failures to use or maintain sedimentation and erosion control devices. This indicates that these devices are functional in preventing intrusion of construction-generated sediments in the absence of major storm events. Particularly in the vicinity of wetlands, silt fences should be erected before any fill is placed on the roadway but these do not substitute for sedimentation traps when subjected to sheet flow.

Culvert construction presents a more difficult problem in that dewatering increases the stream flow and diversion increases the stream velocity. A riprap barrier to reduce outlet velocity is desirable but a sedimentation basin might be also necessary. An increase in salts from culvert construction or other paving operations is also expected and does not appear to increase salts to harmful levels. Changes due to construction are short term. Sediment and turbidity return to normal values within a month while dissolved metals contributing to hardness become normal in six months. The sediment compositions downstream of construction return to preconstruction values within two to three months but aquatic plant life takes longer to be reestablished.

Consideration should be given to scheduling of potentially disruptive operations. The sediment intrusion into Herring Branch had little observed biological effect because it occurred during winter. Such an event might have more consequence during the swamping season of aquatic invertebrates and vertebrates. Hydrologic changes from construction are inevitable but short-term changes appear to have little effect on the threatened plants. However, the effects would be minimal it construction could be concentrated during dormant periods.

Acknowledgements

This study was initiated under project manager Raymond Richter of the Delaware Department of Transportation and continued under the supervision of Joseph Wutka. Sample collection was performed by the authors, Charles Chiu, Yu-Chieh Huang and Mei-Chi Hsu. Charles Chiu and Yu-Cheh Huang performed laboratory analysis and Mei-Chi Hsu did the measurements of total organic carbon. This work was supported by grants from the Delaware Transportation Institute. The views expressed are those of the authors and do not represent the conclusions of the Delaware Department of Transportation.

References

Adams, V. D. (1995). Water and Wastewater Examination Manual, Chelsea. MI, 247pp.

APHA, AWWA, and WPCF (1985). *Standard Methods for Examination of Water and Wastewater*. 16th Ed. 1268 pp.

Bitter, S. D. and K. J., Bowers (1993). "Wetland mitigation and stream restoration." *Public Works*, 124, 50-52.

Burns, J. W. (1972). "Some effects of logging and associated road construction on Northern California streams." *Transactions American Fisheries Society*, 101, 1-17.

Childers, D. L., and J. G., Gosselink (1990). "Assessment of cumulative impacts to water quality in a forested wetland landscape." *Journal of Environmental Quality*, 19, 455-464.

Darnell, R. M., Pequegnat, W. E., Benson, B. J., and Degenbaugh, R. A. (1976). "Impacts of construction activities in wetlands of the United States." EPA-600/3-76-045, USEPA Washington, D. C. 393pp.

Delaware Department of Transportation (1991). U. S. Army Corps of Engineers Section 404 Permit Application package, Contract No. 88-013-01 and 88-013-02. Delaware Department of Transportation, Dover, DE. 41pp.

Delaware Department of Natural Resources and Environmental Control (1993). *State of Delaware Surface Water Quality Standards*, Dover, DE 69pp.

Detenbeck, N. E., P. W., Denver, G. J., Niemi, and A. Lima (1992). "Recovery of temperate-stream fish communities from disturbance: A review of case studies and synthesis of theory." *Environmental management,* 16, 33-53.

Eaglin, G. S., W. A., Hubert (1993). "Effects of logging and roads on substrate and trout in streams of the Medicine bow national forest, Wyoming," *Northern American Fish Management*, 13, 844-846.

Grayson, R. B., S. R. Haydon, M. D. A., Jayasuriya, and B. L Finlayson (1993). "Water quality in mountain ash forest: Separating the impacts of roads from those of logging operation." *Journal of Hydrology*, 150, 459-480.

Harr, R. D. and R. L. Fredriksen (1988). "Water quality after logging small watershed within the Bull Run watershed." *Water Research Bulletin*, AWRA 24, 1103-1111.

Jeglum, J. K. (1975). "Vegetation-habitat changes caused by damming a peatland drainageway in Northern Ontario." *Canadian Field Naturalist*, 89, 400-412.

Jurik, T. W., S. C. Wang, and A. G. van der Valk (1994). "Effects of sediment load on seeding emergence from wetland seed banks." *Wetland*, 14, 159-165.

Lake, D. L. (1987). "Chemical speciation of heavy metals in sewage sludge and related matrices." CRC Press, Boca Raton, FL.

Maryland Department of Transportation (1989). *Erosion and Sediment Control*, (unpaged.)

Ogbeibu, A. E. and R. Victor (1989). "The effects of road and bridge construction on the bank-road macrobenthic invertebrates of a Southern Nigerian stream." *Environmental Pollution*, 56, 85-1000.

Reed, J. R., Jr. (1977). "Stream community response to road construction sediments." Bulletin 97, Virginia Water Resources Research Center, Virginia Polytechnic Institute and State University, 64pp.

Smith, M. E. and J. L. Kaster (1983). "Effect of rural highway runoff on stream benthic macroinveteerbrates." *Environmental Pollution (Series A)*, 32, 157-170.

Stoeckel, J. H. (1967). Wetland road crossing: Drainage problems and timber damage research note Nc-27, U.S. Forest Service, 4pp.

Tessler, A., P., G. C. Campbell, and M. Bisson (1979). "Sequential extraction procedure for the speciation of particulate trace metals." *Analytical Chemistry*, 51, 844-850.

Thibodeau, F. and N. H. Nickerson (1985). "Changes in wetland plant association induced by impoundment and draining." *Biological Conservation*, 33, 269-279.

Thomson, N. R., E. A. McBean, I. B. Mostrenk, and W. J. Snodgrass (1994). "Characterization of stormwater runoff from highways." In *Current Practices in Modeling the Management of Stormwater Impacts* (Ed. W. Jesse), CRC Pres, Boca Raton, FL.

Tikkanen, P., P. Laasonen, T. Muotka, and K. Kuusela (1994). "Short-term recovery of benthos following disturbance from stream habitat rehabilitation." *Hydrobiologia*, 273, 121-130.

U. S. Geological Survey (1994). Water Resources Data for Maryland and Delaware Year 1994. Volume 1. Surface Water Data. U. S. Geological Survey, Washington, D.C. 422p.

Yousef, Y., A. T. Havited-Jacobsen, M. P. Wanielista, and H. H. Harper (1987). "Removal of contaminants in highway runoff following through swales." *Science of the Total Environment*, 59, 391-399.

Review of Constructed Wetlands Applied in Water Pollution Control in Taiwan

Lei Yang[1]

Abstract

Water pollution is a serious problem in Taiwan. Large amounts of pollutants from different point- and nonpoint-sources are discharged into both natural and artificial water bodies, including rivers, lakes, ponds, reservoirs and coastal water. Contaminated water has affected the supply of water in living, industry and agriculture in Taiwan. Wetlands are lands where the water surface is near the ground surface for long enough each year to maintain saturated soil conditions, along with the related vegetation. There are many valuable functions of wetlands including providing habitat for reproduction, feeding, and resting for organisms, flood attenuation, groundwater recharge, sediment entrapment, and water quality purification. Constructed wetlands are thus used as wastewater treatment facilities through the function of water quality purification of wetlands. Besides, this natural treatment system was found more economical than conventional mechanical treatment system for water pollution control. In Taiwan, the research work on constructed wetlands was begun from several years ago conducted by several researchers in the project supported by National Science Council. Different types of water bodies, including contaminated river water, aquaculturing pond water, sewage, industrial wastewater, and storm water, were tested by using microcosm or macrocosm constructed wetland systems for their removal efficiencies of different types of pollutants. The results of this three-year research project showed that the effluents from constructed wetlands exhibited high potential for water reuse, which was beneficial for sustainable use of water resource in Taiwan.

Introduction

Water pollution is a serious environmental problem in Taiwan. In this island most of the rivers and drainage are contaminated by different types of pollutants discharged from different sources of pollution including municipal sewage, industrial wastewater, swine wastewater, and agricultural, storm-water and urban runoff. Due to this problem, the supply of clean water for living, industrial and agricultural purposes was become more and more difficult. Thus, the government of Taiwan plans to develop water resources in the upstream of rivers in order to get clean water for water supply. However, development of new water resources in the mountain areas may destroy the natural ecosystem. Although this way can solve the problem of shortage of clean water temperately, the problem of water pollution may

[1] Professor, Department of Marine Environment and Engineering, National Sun Yat-sen University, Kaohsiung 804, Taiwan

become more serious in the downstream of rivers due to less dilution effect by low flow. Hence, to answer the question of how to solve the problem of shortage of clean water without negative effects, conducting water pollution control is the best answer.

The application of conventional wastewater treatment processes, that mechanical treatment facilities, including sedimentation tanks, aeration basins and filters are installed in a wastewater treatment plant, can certainly achieve the purpose of water pollution control. However, the mechanical treatment processes have disadvantages of complexity of operation, high cost in construction, operation and maintenance, and energy consuming. Natural treatment systems, including land treatment systems, stabilization ponds, aquatic treatment systems, and wetland systems, provide not only low cost of construction, maintenance and operation, and low energy consumed, but an aesthetic view and ecological habitat for wild life. Thus, more and more countries use natural systems as an alternative for water pollution control. The removal mechanisms of pollutants in natural systems include physical, chemical and biological processes occurred naturally by natural components of non-living stratum, such as soil and gravel, and living organisms, such as bacteria and aquatic plants.

Wetlands are lands where the water surface is near the ground surface for long enough each year to maintain saturated soil conditions, along with the related vegetation (Reed *et al.*, 1995). Wetlands are among the most important components in natural ecosystems. The important functions and values of wetlands include flood water storage, flodding reduction, erosion control, sediment stabilization, sediment/toxicant retention, groundwater recharge/discharge, natural water purification, nutrient removal/retention, chemical and nutrient sorption, food chain support, fish and wildlife habitat, migratory waterfowl usage, recreation and uniqueness/heritage values (Hammer and Bastian, 1989; Day *et al.*, 1998; Hughes *et al.*, 1998; Mitsch and Cronk, 1995; Hamilton *et al.*, 1997; Corbitt and Bowen, 1994). These functions and values have obvious natural benefits but also identify how wetlands may be used as an effective mechanism for wastewater treatment, either in their natural state or as a constructed system. Constructed wetlands are thus used as wastewater treatment facilities through the function of water quality purification of wetlands. Hammer and Bastian (1989) defined constructed wetlands as a designed and man-made complex of saturated substrates, emergent and submergent vegetation, animal life, and water that simulates natural wetlands for human use and benefits. Some synonymous terms including man-made, engineered, artificial and treatment wetlands are also used. Although bogs and swamps have been constructed or used for wastewater treatment, most constructed wetlands for wastewater treatment are preferred to be in the form of marshes for the reason of having the most promise for wastewater treatment efficiency (Hammer and Bastian, 1989). Much of the wastewater treatment in a constructed wetland occurs by means of bacterial metabolism and physical sedimentation, as it does in an activated sludge or trickling filter treatment facilities. Wetland plants just serve to provide structure to support the algae and bacterial population that provide wastewater treatment capability and reliability in a created wetland community environment (Corbitt and Bowen, 1994). However, through absorption and assimilation, wetland plants can remove nutrients from water and wastewater for biomass production (growth). Besides, one abundant by-product of the plant growth process is oxygen, which can increase the dissolved oxygen (DO) content of the water and also of the soil in the immediate vicinity of plant roots. This process is called as "root zone effect (RZE)" and shown in Figure 1 (Hammer and Bastian, 1989). The wetland plants commonly used in constructed wetlands include water hyacinth, duck weed, water lily (submerged plants), cattail, reed, and bulrush (emergent plants) as shown in Figure 2.

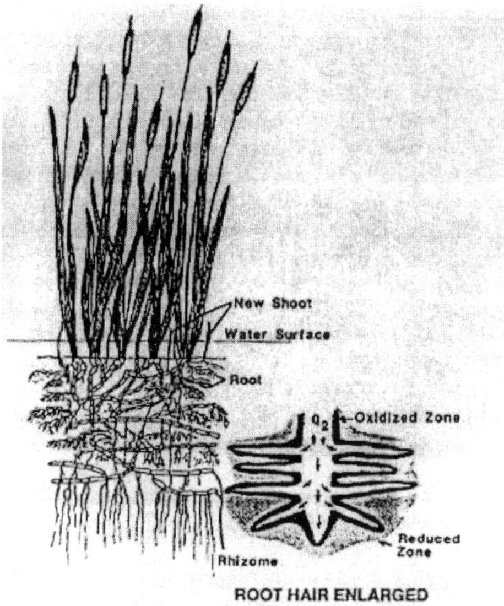

Figure 1. Wetlands plants have the unique ability to transport oxygen to support their roots growing in ananerobic substrates. (Hammer and Bastian, 1989)

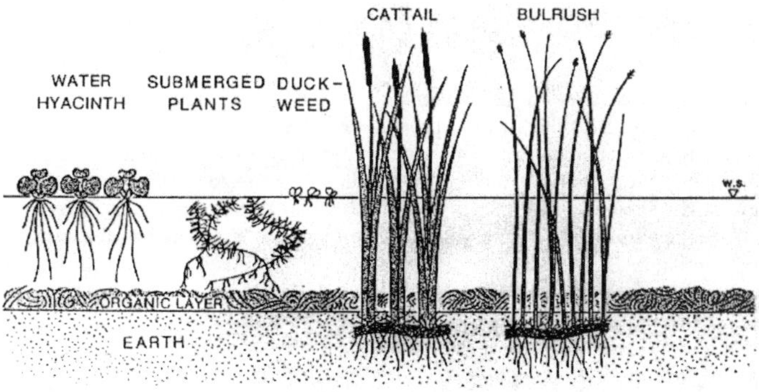

Figure 2. Common aquatic plants in constructed wetland systems. (USEPA, 1991)

Constructed wetlands have been widely applied in treating different types of wastewater, such as sewage (Reed *et al.*, 1988, Yang *et al.*, 2001; Yang, 2001), storm water (Green and Martin, 1996), polluted river water (Jing[a] *et al.*, 2001; Jing[b] *et al.*, 2001), industrial wastewater (Vrhovsek *et al.*,1996), agricultural runoff (Higgins *et al.*, 1993), acid mine drainage (Karathanasis and Thompson, 1995) and landfill leachate (Bernard and Lauve, 1995). Such natural treatment systems have been shown to have a significant capacity for both wastewater treatment and resource recovery (Hofmann, 1996). According to the flow patterns and media applied, constructed wetlands are divided into two systems: subsurface flow (SF) and free water surface (FWS). In SF systems, wastewater flows underneath through the plant-rooting media (gravel mostly), while flow over the surface of the rooting media (usually soil) in FWS systems. Since the water surface in SF systems is maintained below the tip of the gravel media, there are less problems arisen from odors, insect vectors, or public exposure, which are thus suitable to treat landfill leachate, farm wastewater and some industrial wastewater. However, FWS systems have advantages of lower installation cost and simpler hydraulics making them still used widely in treating polluted river or lake water for restoration, storm water runoff from agricultural lands and secondarily or tertiary treated sewage effluent (Kruzic and White, 1996). For both of the systems, aquatic vegetation has been reported to transport approximately 90% of the oxygen available in the rhizophere (Reddy *et al.*, 1989). Thus, it creates aerobic conditions in the root zone of plants (i.e. root zone effect mentioned previously) to help to develop both heterotrophic and autotrophic microorganisms, which participate in the aerobic decomposition of organic matter and nitrification, respectively (Brix, 1993). Nitrogenous (N) and phosphorous (P) nutrients are two major pollutants considered for removal in wetland systems. Removal processes for N include uptake by plants and other living organisms, ammonification (mineralization), nitrification, denitrification, ammonia volatilization, and cation exchange for ammonium (Brix, 1993). As to the removal mechanisms for P occurred in wetlands, they include chemical adsorption and precipitation and biological transformations (Sakadevan and Bavor, 1998). However, Adler and his research fellows (1996) have reported that the main nutrient removal mechanisms in wetlands are biological denitrification of nitrogen by microbes and chemical adsorption of phosphorus on hydrous oxides of Fe, Al and Ca minerals in soils. Generally, constructed wetland system, which was one of the natural treatment systems, was found more economical than conventional mechanical treatment system for water pollution control.

In Taiwan, the research work on constructed wetlands was begun from several years ago conducted by several researchers in the project supported by the National Science Council. Different types of water bodies, including contaminated river water, aquaculturing pond water, sewage, industrial wastewater, and storm water, were tested by using microcosm or macrocosm constructed wetland systems for their removal efficiencies of different types of pollutants. The results of this three-year research project showed that the effluents from constructed wetlands exhibited high potential for water reuse, which was beneficial for sustainable use of water resource in Taiwan. We will review each of these researches in the following sections.

Constructed Wetlands Applied in Industrial Wastewaters Treatment

The effluents from wetland treatment systems may be reused as recirculated water to achieve the sustainable use of water resources. The types of wastewater used in constructed wetland treatment systems include agricultural runoff, sewage, stormwater drainage, landfill leachates, mine drainage, and industrial wastewater. Since constructed wetlands offer cost-effective treatment potential, use of this type of natural systems replacing traditional mechanical ones for wastewater treatments become more and more prevailing in Europe, America, Australia, and Mainland China. Due to lack of suitable lands for constructed wetlands, in Taiwan, there are no authorities, including government and private companies, considering to use constructed wetlands to treat their wastewater. However, owning to the same reasons, more and more industrial parks will be developed through land reclaim by dredging in the coastal areas of Taiwan. Hence, the problems of lack of lands will be solved. Therefore, we can consider using constructed wetlands as advanced treatment systems for the wastewater produced from those coastal industries. Due to the lack of fresh water resources in coastal zones, the effluents from wetlands might be recycled and reused in the water supply systems of coastal industries to sustain the water resources. Thus, the feasibility of using the effluents from constructed wetlands treating partially treated industrial wastewater as recirculated and reused water for those industries located in coastal zones was studied (Yang, 2000). In the meanwhile, the complement between constructed wetlands and ecological conservation will also be studied.

In Yang's study (2000), based on the results achieved in the first year's study that used lab-scale constructed wetlands to test, we discussed the treatment efficiencies of oil-refinery industry wastewater by pilot-scale constructed wetland. The purpose this study is to decrease certain pollutants in the effluents investigate the feasibility of water reuse and recovering. The constructed wetland was a free water surface system (FWS) filled with the sandy media, while was planted with *Phragmites communis*. The flow rate of inflow for the treatment wetland was controlled at 330 mL / min. As a result, orthophosphate (PO_4^--P) and TSS were reduced over 80% . Also, TP, COD, NH_3 and TKN were reduced above 60 % .Oil and grease were reduced about 20% . However, the removal efficiencies of nitrite (NO_2^-) and nitrate (NO_3^-) were poor.

Constructed Wetlands Applied in Sewage Treatment

In the study by Kuo (2000), constructed wetland was built next to a secondary wastewater treatment plant in the National Pingtung University of Science and Technology to polish the effluent intended for water reclamation. Three units of the same dimension (L×W×H) 10m×3m×1.2m were built, with three different media: air-cooled blast furnace slag, rocks, and artificial fiber media. The Hydraulic Retention Times (HRTs) tested were 1/6 day, 1 day, 3 days, and 5 days. The Organic Loading Rates (OLRs) tested dependent on the influent concentrations of BOD_5 or COD, was found to between.

In the overall performances in slag filter bed and rock filter bed, removal efficiencies of the Suspended Solids (SS) were 64%±17(n=24) and 53%±26(n=24); Total-COD were 31%±20(n=24) and 31%±12(n=24); Soluble-COD were 34%±16(n=15) and 36%±17(n=15); TKN were 19%±11(n=14) and 21%±15(n=14); NH_3-N were 22%±13(n=18) and 28%±29(n=18); NO_3-N were 50%±12(n=19) and 60%±34(n=17); Total Phosphorous (TP) were 39%±20(n=19) and 17%±16(n=19); and the removal efficiencies of PO_4-P were 36%±25(n=19) and 21%±21(n=19), respectively.

In Kuo's study (2000), slag filter bed showed better performance than rock filter bed in SS and BOD removal. This could be attributed to the reasons that slag can provide more surface area due to the small holes on the surface, microorganisms could attach on the surface area, and SS could settle better. The plants grown in slag filter bed were not as good as the plants grown in the rock filter bed. It is possible that some unidentified elements in the slag could inhibit the growth of plants. It's important to find out the reason before the slag is to be applied in constructed wetland.

In another study of sewage treatment by constructed wetland systems in Taiwan (Yang *et al.*, 2001), four lab-scale microcosm constructed wetlands, including gravel-beds with plant (A1) and without plant (A2), and soil-beds with plant (A3) and without plant (A4), were used to conduct the nutrient removal tests. The influent used in the tests is primarily treated sewage, while the plant selected is Napier grass (*Pennisetum purpureum*). The purpose of this study is to compare the removal efficiencies of nitrogenous and phosphorus nutrients among these four microcosm wetland systems based on statistical analyses. Three factors, namely, with/without vegetation (F_1), medium types of gravel/soil (F_2), and time period for the test run of first/second stage (F_3), and four combined effects of factors, F_1 by F_2 ($F_{1,2}$), F_1 by F_3 ($F_{1,3}$), F_2 by F_3 ($F_{2,3}$), and F_1 by F_2 by F_3 ($F_{1,2,3}$), were run by an ANOVA model to analyze the relationships between the amounts of nutrient removed from the wetland systems and these seven factors. According to the analytical result, we found that the removals of ammonia (NH_3-N), nitrate (NO_3^--N), and soluble reactive phosphorus (SRP) were related to these factors and combined effects of the factors. It was found that the main removal mechanism for NH_3–N was nitrification, which could be enhanced by the root zone effect (RZE) in the vegetated gravel-bed wetland systems, while NO_3^--N were removed mainly by denitrification and plant uptake in vegetated systems. However, the main removal mechanism for SRP was chemical adsorption in the unsaturated soil-bed systems. The effect of plant litter was also a significant mechanism affecting nutrient removal in the surface flow pattern soil-bed wetland systems without harvest.

In the study conducted by Yang (2001), two lab-scale, two-stage constructed wetland systems, including gravel-soil beds with or without plants were used to study the nutrient removal effects of primarily treated sewage. According to his experimental results, in both of the systems, the gravel beds (the first stage) removed most of the N and P nutrients (above 70% of removal efficiencies) during the three-month running time after the systems had reached a steady state. However, the soil bed (the second stage) in the vegetated wetland system showed little increase in N nutrient, while further removal of P nutrient was found in the bed. In the system without vegetation, we found that both N and P nutrients were removed further in the soil bed. The reason might be that the plant litter was accumulated on the soil surface of the vegetated bed with a free water surface flow regime, in which the plant uptake effect might be less than plant litter release effect on N nutrient. However, the accumulation of plant litter did not affect the nutrient levels in the effluent of the vegetated gravel bed due to its subsurface flow pattern. The average first-order rate constants for nitrogen removal in the systems with and without vegetation were measured to be equal to 0.067 and 0.073 day^{-1}, respectively, while the numbers were equal to 0.045 and 0.107 day^{-1}, respectively for phosphor removal. The conclusion is that the biological processes, including microflora processing and macrophyte uptake, were dominant in the vegetated gravel bed to remove N nutrient, while P nutrient was removed mainly by chemical adsorption in the soil bed with or without plant. It might be that (due to) the higher porosity and significant root zone effects exist in the vegetated gravel bed system and that (to) the mineral clay soil type has low porosity in the soil bed.

Constructed Wetland Applied in Aquaculturing Wastewater Treatment

In Taiwan, Lin and Jin (2000) have studied the effects of using constructed wetland systems to treat aquaculturing wastewater for the purpose of recirculation. In their study, a pilot-scale constructed wetlands system consisting of a free water surface (FWS) wetland and a subsurface flow (SSF) wetland run in series was set up for treating aquaculture farm wastewater in this study. The FWS wetland was planted with water spinach (*Ipomoea aquatica*) in the front half and with a native weed (*Paspalum vaginatum*) in the second half, and the SSF was planted with common reed (*Phragmites communis*). The first part of this study was conducted to examine the system performance in removing suspended solids (SS), algae, chemical oxygen demand (COD), inorganic nitrogen and phosphate under various hydraulic loading rates (1.8 to13.5 cm d^{-1}). Several significant findings were obtained in this study, which included: (1) the start-up periods for achieving stable performance of pollutants removal and macrophyte establishment; (2) regression equations for correlating the effluent qualities and the loading rates; (3) the relationship between pollutants removal rate and loading rate; (4) first-order removal rate constants for various pollutants. These results could be provided as basic information for process design and operation in constructed wetlands system for aquaculture wastewater treatment. In the second part of this study, we set up a recirculating aquaculture system (RAS) that consisted of a tank for fish rearing and a SSF wetland for recirculated water treatment. A control aquaculture system (CAS) containing the same size of the rearing tank as the RAS, but without water recycling, was also run for comparison. A culture density of 7 fish m^{-2} of milkfish and 61 prawn m^{-2} of Thailand shrimp was employed in the rearing tanks of both systems. The performance of the SSF wetland and the water qualities and the culture species growth in the two rearing tanks were investigated. The recirculated water was performed at a rate of around 15 % of the water volume of the rearing tank per day. During the 12 weeks experimental duration, SSF wetland showed efficient removal for BOD_5, nitrite, and microorganisms (algae, total coliform, *Vibrio*) from the recirculated water (> 70%), while maitaining substantial removal for SS, ammonium, nitrate, and phosphate (30-60 %). The water qualities of SS, nitrate, algae, turbidity, total coliform, and *Vibrio* in the RAS rearing tank were significantly lower ($p < 0.33$) than those in the CAS rearing tank, whereas the levels of BOD_5, ammonium, and nitrite in the rearing tanks between the two systems were not significantly different ($p > 0.7$). The body length of the culture species increased 272 for milkfish and 1510 % for Thailand shrimp in the RAS system, which was 26 and 250 % higher than those in the CAS ($p = 0.08$ and < 0.005).

Application of Constructed Wetlands in Polluted River Water

The Erh-Ren River is one the most polluted rivers in Taiwan. Although its flow rate is relatively low, the rate is still beyond the capacity of any traditional water treatment facility. A pilot-scale constructed wetland (CW) was the attempt used to purify the highly polluted river water and to collect data for the construction and operation of a full-scale system in the future (Jing[a] *et al.*, 2001). During the study, the most efficient nutrient removal occurred between April and October. The monthly average removal rates for chemical oxygen demand (COD), ammona-N (AN) and orthophosphate (OP) were ranged from 13-51%, 78-100%, 52-85%, respectively. After November, imput COD levels increased, and the monthly average removal rates of AN dropped to 26%, while OP was decrease to 13%. The dramatic changes in removal efficiency suggest that the macrophytes in the CW had a direct influence on the water treatment and that the change of seasons and the quality of the river water inhibited the growth of the macrophytes.

In another study on Erh-Ren River (Jing[b] *et al.*, 2001), a pilot-scale constructed wetland (ERRCW) system containing a free surface flow wetland (FSF) followed by a subsurface flow wetland (SSF) was set up to treat the polluted Erh-Ren River water. The system was designed to include solids removal. A removal efficiency of suspended solids (SS) in the ERRCW system that was greater than in a control system indicated that the existence of macrophytes in wetlands cold enhance solids removal. The results of this study also showed that SS were removed mainly through the SSF in the ERRCW system. The poor removal of SS, or even an increase in SS, that occurred in the FSF was found to be caused by the growth of algae. Although the macrophytes reduced the penetration of sunlight into the water in FSF portion of the ERRCW system, the residue of nutrients in water and a low flushing rate still induced the growth of algae on some open water surface. From this study, it may be concluded that the roles played by macrophytes in solids removal include providing barriers for filtration, reducing sunlight penetration, supplying oxygen to enhance biodegradation of organic solids, and absorbing nutrients to limit the growth of algae.

Wetland Systems Applied in Non-point Source Pollution Control

In Taiwan, non-point source (NPS) pollution is one of the major causes of impairment of surface waters. The non-point source pollution controlled by wetland systems has been studied by Kao (2000). In his study, the constructed wetland systems were built inside the campus of National Sun Yat-Sen University to remove (1) NPS pollutants due to the stormwater runoff, and (2) part of the untreated wastewater from school drains. The plants grown on the wetland included floating (*Pistia stratiotes* L.) and emergent (*Phragmites communis* L.) species. Analytical results indicated that the constructed wetland removed significant amount of NPS pollutants and wastewater constituents. Results from this study would be very useful in the design of constructed wetlands for NPS pollutin control and water quality improvement.

Fate of the Pollute in the Wetland System

The fate of pollute in wetland systems has also been studied in Taiwan (Wang and Lee, 2000). The main goal of this research was to study the relation between various metal (Pb, Cr, Zn) removal efficiencies and macrophyte growth in constructed wetlands (CW) using fixed influent concentration (20 mg/l). The artificial wastewater also contained 5 mg P/l orthophosphate and 25 mg N/l ammonia. The CW's were free water surface systems and four systems were operated in parallel. Each system planted with different macrophytes (*Phragmites communis L., Pistia stratiotes L., Commelina communis L.,* and vetiver) with hydraulic retention time 4 days (flow rate = 2.2ml/min). In these systems, removal efficiencies of Pb were higher than 94% of Cr ranged from 75 to 87% and of Zn were lower than 40%. The Pb contains in plant tissues were all increased after treatment indicating plant accumulation; Cr contains in plant tissue were decreased probably due to dilution because of plant growth; however, Zn contains were remained low before and after treatment. Although metal concentrations in roots of each system were the highest amount various tissues, the total accumulation of metal was found in the leaves. In the whole plant, the accumulation of Pb ranged from 22 to 95 mg; Cr ranged from 64 and 352 mg; and Zn ranged from 3 to 12 mg. The Pb accumulations in soil ranged from 1172 to 1660 mg; Cr ranged from 118 to 4425 mg;

and Zn ranged from 137 to 1277 mg. Therefore, soil sorption was the major sink for metal removal in a CW system.

Development of Database of Designing Constructed Wetland in Taiwan

Li (2000) has localized the design criteria of constructed wetland systems for wastewater treatments. His study presented the performance of constructed wetland for improving water quality by using pilot scale constructed wetlands, consisting of two cells of subsurface wetland and two cells of free surface wetland. There are two types of influent sources to the constructed wetland – the discharged wastewater from a secondary wastewater treatment plant and the lake water. Based on the result, the treatment performance is good for removing nutrients from the secondary effluent. But the suspended solid, volatilized suspended solid, and the total organic carbon, are somewhat poor. Similarly, the treatment performance for lake water is good in the nitrate removal, but poor in volatilized suspended solid, biological oxygen demand, and the total organic carbon. The subsurface flow constructed wetland can remove phosphorus, total phosphorus, in contrast, the free constructed wetland is incapable of removing total phosphorus.

The obtained experiment data were modeled using the first order kinetic theory. In comparison with the reported results, the kinetic coefficients(k)of suspended solid is low, but the value of k for the other pollutants is close to that reported. We also used visual basic for application (VBA) to develop a computer program for designing constructed wetland. Through the input of the flow rate, the influent concentration, and the expected effluent concentration as well as kinetic coefficient k, the constructed wetland area can be determined. Though the applicability of the program has been demonstrated in personal computer, its reliability needs further test.

Conclusions

Since Taiwan is located in subtropical areas, it is very suitable to use constructed wetland systems to treat the contaminated water in Taiwan. However, according to the experimental results of those studied conducted in Taiwan, even though constructed wetlands have been proved performing well for water pollution control, there is still no any field-scale constructed wetlands built or under construction in Taiwan to treat the polluted water bodies. The reasons may be summarized as followed:

1. The EPA officials of Taiwan think that there is no enough land provided for building constructed wetlands in Taiwan.

2. Wetlands may be the sites to raise mosquito, which are the media causing some disease.

3. Wastewater is still needed pretreatment before it is discharged into constructed wetland systems, which still cost more in treating wastewater.

4. There is no legislation on the issue of wetland systems in Taiwan.

Therefore, first of all we should make more efforts on pushing the legislation work on protection, conservation and creation of wetlands in Taiwan. Then, building field-scale constructed wetland systems for water pollution control in Taiwan may become possible. In the future, the sustainability of water resources will thus be achieved in Taiwan.

References

Adler, P. R., S. T. Summerfelt, D. M. Glenn., and F. Takeda (1996). "Evaluation of a wetland system designed to meet stringent phosphorus discharge requirements." *Water Environment Research*, 68, 836-840.

Bernard, J. M. and T. E. Lauve (1995). "A comparison of growth and nutrient uptake in *Phalaris arundinacea L.* growing in a wetland and a constructed bed receiving landfill leachate." *Wetlands*, 15, 176-182.

Brix, H. (1993). "Wastewater treatment in constructed wetlands: system design, removal processes, and treatment performance." *Constructed Wetlands for Water Quality Improvement*, G. A. Moshiri (ed.), Lewis Publishers, Boca Raton, FL., 9-22.

Corbitt, R. A. and P. T. Bowen (1994). "Conbstructed wetlands for wastewater treatment." *Applied Wetlands Science and Technology*. D. M. Kent (ed.), Lewis Publishers, Inc., Ann Arbor, Michigan, 221-242.

Day, J. W., Jr., A. Rismondo, F. Scarton, A. Are, and G. Cecconi (1998). "Relative sea level rise and venice lagoon wetlands." *Journal of Coastal Conservation*, 4(1), 27-34.

USEPA (1991). "Design manual of constructed wetlands and Aquatic plant systems for municipal wastewater treatment." Office of Research and Development, U.S. Environmental Protection Agency, EPA/625/1-88/022.

Green, M. B. and J. R. Martin (1996). "Constructed reed beds clean up storm overflows on small wastewater treatment works." *Water Environment Research*, 68, 1054-1060.

Hamilton, S. K., S. J. Sopple, D. F. Calheiros, and J. M. Melack (1997). "An anoxic event and other biogeochemical effects of the pantanal wetland on the Paraguay River." *Limnology and Oceanography*, 42(2), 257-272.

Hammer, D. A. and R. K. Bastian (1989). "Wetland ecosystems: natural water purifier?" *Constructed Wetlands for Wastewater Treatment-Municipal, Industrial and Agricultural.* D. A. Hammer (ed.), Lewis Publishers, Inc., Ann Arbor, Michigan, 6-20.

Higgins, M. J., C. A. Rock, R. Bouchard, and B. Wengrezynek (1993). "Controlling agricultural runoff by use of constructed wetlands." *Constructed Wetlands for Water Quality Improvement*, G. A. Moshiri (ed.), Lewis Publishers, Boca Raton, FL., 359-367.

Hofmann, K. (1996). "The role of plants in subsurface flow constructed wetlands." *Ecological Enginneering for WastewaterTtreatment*, C. Etnier and B. Guterstam (ed.), Lewis Publishers, Boca Raton, FL., 183-196.

Hughes, C. E., P. binning, and G. R. Willgoose (1998). "Characterisation of the hydrology of an estaurine wetland." *Journal of Hydrology* (Amsterdam), 211 (1-4), 34-49.

Jing[a], S. R., Y. F. Lin, D. Y. Lee, and T. W. Wang (2001), "Using constructed wetland systems to remove solids from highly polluted river water." *Water Science Technology: Water Supply*, 1(1), 89-90.

Jing[b], S. R., Y. F. Lin, D. Y. Lee, and T. W. Wang (2001). "Nutrient removal from polluted river water by using constructed wetlands." *Bioresource Technology*, 76(2), 131-135.

Kao, C. M. (2000). "Surveying and study of hydrological phenomena and removal of pollutants from non-point sources in natural wetlands." *Abstract Proceeding of Conference of Sustainable Development in Taiwan*, Division of Sustainable Development, National Science Council, Taipei, Taiwan, 126.

Karathanasis, A. D. and Y. L. Thompson (1995). "Mineralogy of iron precipitates in a constructed acid mine drainage wetland." *Soil Science Society of American Journal*, 59, 1773-1779.

Kruzic, A. P. and K. D. White (1996). "Natural treatment and on-site processes." *Water Environment Research*, 68, 498-502.

Kuo, W. C. (2000). "Study of constructed wetlands for sewage treatment and reuse." *Abstract Proceeding of Conference of Sustainable Development in Taiwan*, Division of Sustainable Development, National Science Council, Taipei, Taiwan, 122-123.

Li, C. Y. (2000). "Development of databas of designing constructed wetland in Taiwan." *Abstract Proceeding of Conference of Sustainable Development in Taiwan*, Division of Sustainable Development, National Science Council, Taipei, Taiwan, 128-131.

Lin, Y. F. and S. R. Jin (2000). "Application of constructed wetlands in recirculation treatment of aquaculturing wastewater." *Abstract Proceeding of Conference of Sustainable Development in Taiwan*, Division of Sustainable Development, National Science Council, Taipei, Taiwan, 124-125.

Mitsch, W. J. and J. K. Cronk (1995). "Influence of hydrologic loading rate on phosphorus retention and ecosystem productivity in created wetland." Final Report WES/TR/WRP Re-6.

Reddy, K. R., E. M. D'Angelo, and T. A. DeBusk (1989). "Oxygen transport through aquatic macrophytes: the role in wastewater treatment." *Journal of Environmental Quality*, 19, 261-270.

Reed, S. C., E. J. Middlebrooks, and R. W. Crites (1988). *Natural Systems for Waste Management and Ttreatment*. McGraw-Hill, New York, 221-256.

Reed, S. C., R. W. Crites and E. J. Middlebrooks (1995). "Wetland systems." *Natural Systems for Waste Management and Treatment*, 2nd Ed., McGraw-Hill, Inc., New York, 173-284.

Sakadevan, K. and H. J. Bavor (1998). "Phosphate adsorption characteristics of soils, slags and zeolite to be used as substrates in constructed wetland systems." *Water Research*, 23, 393-399.

Vrhovsek, D., V. Kukanja, and T. Bulc (1996). "Constructed wetland (CW) for industrial waste water treatment." *Water Research*, 30, 2287-2292.

Wang, T. W. and D. Y. Lee. (2000). "Study of fate of the pollute in the wetland system." *Abstract Proceeding of Conference of Sustainable Development in Taiwan*, Division of Sustainable Development, National Science Council, Taipei, Taiwan, 127.

Yang, L. (2000). "Application of constructed wetlands in recirculation treatment of industrial wastewater in coastal zones." *Abstract Proceeding of Conference of Sustainable Development in Taiwan*, Division of Sustainable Development, National Science Council, Taipei, Taiwan, 120-121.

Yang, L., H. T. Chang, and M. N. L. Huang (2001). "Nutrient removal in gravel- and soil-based wetland microcosms with and without vegetation", *Ecological Engineering*, 18(1), 91-105.

Yang, L. (2001). "Nutrient removal from sewage by two-stage constructed wetland systems." *Journal of the Chinese Institute of Environmental Engineering*, 11(2), 137-142.

Apply Neural Networks to Watershed Management

Nian She[1]

Abstract

This paper demonstrates neural network model can be used as a powerful tool in watershed management. Two applications were presented: (1) development and evaluation of several hydrological interpolation schemes for correcting missing or noisy data inflow data in Seattle Public Utility's Tolt River watershed and (2) forecasting short-term water demand in summer months. Neural network model was compared with traditional method such as linear regression. It was found that neural network model performed better than the traditional statistical models, especially when relationship among parameters is not linear.

Introduction

Water supply, demand and quality are major concerns of Seattle Public Utilities. Since the beginning of late 80's the population in metropolitan of Seattle has grew significantly and the growth is projected to continue at the rate of 1.0% annually for the next 20 years (Skeel, et al., 1998). Without developing new water resources the management of existing water supply and quality to meet the regional growth become critical task to Seattle Public Utilities, especially during the summer months when the demand increasing significantly while precipitation is very low. Three issues related to the watershed management are

(1) Yield prediction, which is dependent on the in-stream flow in Seattle Public Utility's Cedar River and Tolt River watershed
(2) daily consumption, which may affect the reservoir operations during the summer months; and
(3) the water quality in the distribution reservoirs.

In issue (1) the key question is how to estimate the data gaps in the in-stream flow? In issue (2) the key question is whether or not we can forecast water consumption in a short-term period, say a week ahead; and the key question issue (3) is that can we adequately model the dynamics of water quality parameters. In this paper I will only answer the first two questions by applying neural networks. The third question will be addressed in a separate paper.

---

[1] Project Manager, Seattle Public Utilities, 720 Second Ave., Dexter Horton Building, Seattle, WA 89104, U. S. A.

The Architecture, Train, Validation and Test of Neural Network

Neural network has been applied to water treatment, water quality and waste water management in recent years (Garrett, et al., 1997; Yabunaka et al., 1997; Baxter, et al., 2001 and Brion et al., 2001). In this paper a neural network with feed-forward architecture and backpropagation learning algorithm was chosen for the tasks. The feed-forward neural network is simple and counts for more than 90% applications in civil and environmental engineering fields. Figure 1 shows a layout for a typical three-layered feed-forward neural network, which contains an input layer made up L input nodes (variables), a hidden layer, made up m hidden nodes, and an output layer, made up n output nodes (variables). The architecture is denoted as a L-m-n neural network. In general, a feed-forward neural network contains an input layer, an output layer, and one or more hidden layers between the input and the output layer. Each node at current layer is connected with nodes in the previous layer by a linear combination of weights and bias (see Figure 1) except for input nodes. An activation function, usually in form of $1/(1+e^{-x})$ or tanh(x), then is invoked to process the combination before sending its output directly to nodes in the next layer except for output nodes.

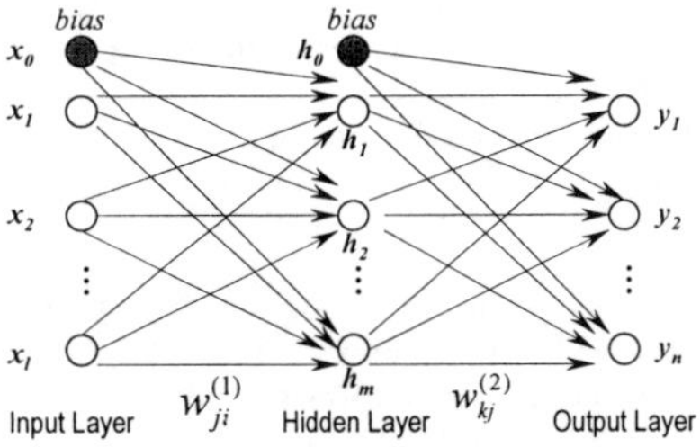

Figure 1. A three-layered feed-forward neural network

After the setup of a feed-forward network the backpropagation algorithm is used to train the network. The backpropagation algorithm uses a gradient descent method to minimize errors. The training process can be viewed as updating network architecture and connection weights and bias so that a network can learn complex systems. That is given a set of input variables and corresponding output variables, the network organizes itself internally so it predicts the desire output. The training process (backpropagation) is an iterative process that has three steps:

(1) starting with a randomized weights and bias, the input variables are forward propagated through the network to yield an output
(2) the output then backpropagates the errors back to the sensitivity according to generalized delta rule (GDR) or other learning rules
(3) the weights and bias are updated using this backpropagated error and learning rules. The process is repeated until the error is within a tolerable limit or range

The detailed steps can be found in many standard textbooks (Hagan, 1996).

In order to ensure that the network is not simply memorizing the training pattern, the network must be validated. When the number of training examples is small, the leave-one-out method can be used to estimate the performance of the network. In this method, the network is trained on all examples but one and then test on the one left out. For N examples, the process is repeated N times leaving out a different example each time. The error statistics on these N tests can be used to determine the performance of the network. When the number of training examples is large, the method can be extended by dividing the examples into N groups. The network then is trained on N-1 groups leaving out a different group each time, and test on the group that is left out. In practice, it is common to divide the examples to two groups, then swap the training and testing groups. This is also called cross-validation technique (Schwek, 1996).

Application 1: missing data estimation/interpolation

As mentioned earlier, Seattle Public Utility's yield forecasting was developed using 65 years of historical daily in-stream flow from US Geological Survey (USGS) gauges in Cedar and Tolt River watershed. Since the records of gauged flow contain many gaps where either there were no gauges before a given time or the existing gauges ceased recording during a time period, those gaps must be filled by estimation or interpolation accordingly. Traditionally, linear regression was used to fill in periods of missing data at existing gauges, i.e., taking X as recorded flow in a nearby gauge, Y as estimated flow in the gauge that has some gaps. If the correlation between the two gauges is strong, the estimation usually is acceptable. However, the regression usually over or under fits extreme values, which is critical in the yield forecasting, especially during the summer months. Instead, neural network can be trained to fit the extreme values better. This can be seen from the following example.

USGS gauge 12148500 is located at up-stream of main stem of the Tolt River where the South Fork Tolt River and North Fork Tolt River merge together (Figure 2).
The gauge has a fairly good record since 1963, but it did not record for period of 1932 – 1937. In order to fill in missing data for the period, fourteen years of overlapping data (1930, 1931 and 1938 – 1949) in the nearby gauge 12142000 was chosen by a consultant firm to regress against gauge 12148500. The regressions were performed for each month. The R^2 has a range of 0.84 to 0.95. An unusually high flow of 6340 cfs recorded on October 31, 1943 at gauge 12148500 was considered as an outlier and eliminated for better fit by the consultant firm. Though the correlation between these two gauges is very strong, the regression does not fit extreme outliers very well.

To demonstrate neural network can improve fitting extreme outliers, October's flow data was chosen because it has a lower R^2 value (0.84). The network architecture is rather simple in this case. The flow recorded at USGS gauge 12142000 and the calendar dates are input variables; the flow recorded at USGS gauge 12148500 is output variable. In this study, both input and output variables are normalized to a unit square with origin (0,0). A typical three layer feed-forward network is sufficient to be trained to learn the pattern. Before training the network, the data was divided to 14 groups naturally since 14 years October's data was compiled sequentially.

Without discard the outliers, a neural network with a 1-5-1 architecture was trained, validated and tested according to leave-one-out method mentioned in previous section. The R^2 for training set ranged from 0.89 to 0.92. The R^2 for testing set ranged from 0.72 to 0.83. Figure 3 shows the validation of 1938 data for linear regression vs. neural network. It is clear that the neural network model fits the peak flow better than linear regression though their R^2 is almost the same (0.72).

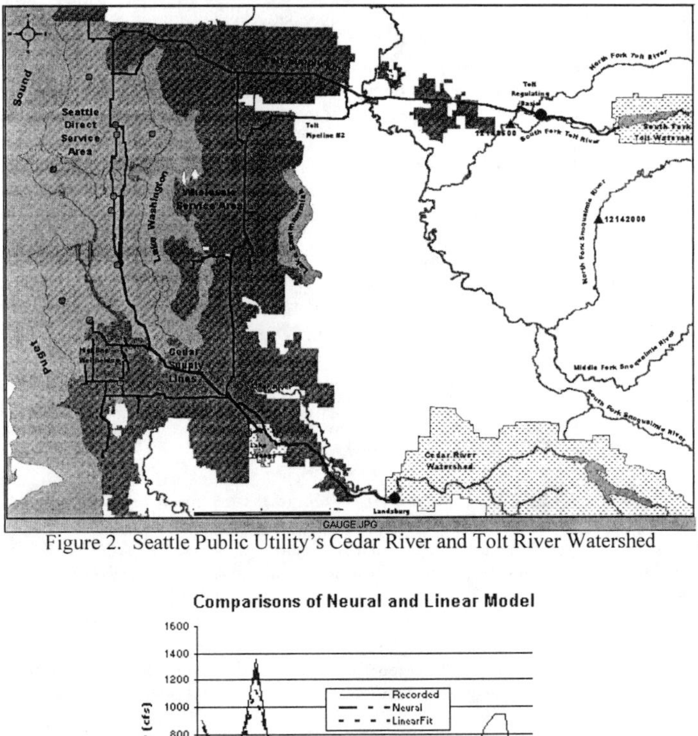

Figure 2. Seattle Public Utility's Cedar River and Tolt River Watershed

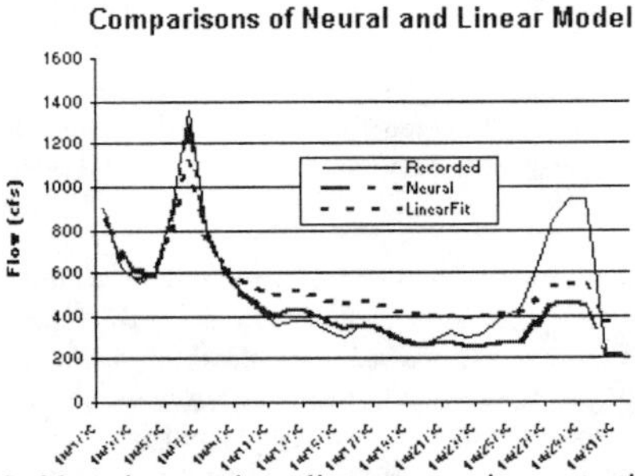

Figure 3. Neural network vs. linear regression on testing data set

Application 2: Forecasting Short-term Water Consumption

Seattle Public Utilities supplies water to 1.3 million people and businesses in the metropolitan of Seattle. Nearly all this water is from the 90,000-acre Cedar River Watershed and the 13,300-acre South Fork Tolt River Watershed in eastern King County (see Figure 1). During the summer months the water consumption increases dramatically. Due to lack of precipitation from June to September, the day-to-day operation of Cedar River and Tolt River reservoirs becomes very important. The operators wish to forecast water consumption a week ahead for better planning and scheduling of maintenance. Unlike the winter months the water consumption varies from 120 to 300 mgd during the summer. The 7-day moving average used by the management is unable to forecast consumption a week ahead with acceptable accuracy. Hence, the neural network model was proposed as an alternative method in year 2000 to forecast summer water consumption.

As mentioned in previous section, a neural network contains an input layer, one or more hidden layer and an output layer. The output variable is water consumption. The input variables include maximum temperature, precipitation, month and indicator of weekday or weekend. Data collected from 1989 to 1999 was investigated. Since the Seattle Public Utility has actively implemented the water conservation plan after 1992's severe drought, the consumption pattern has been changed since 1994. Therefore, the time series data from 1994 to 1999 was used to train the neural network model and data collected in 2000 was used to validate the model.

To determine the reasonable amount of input variables, simple statistical analysis was applied to the rural time series data. One can see from Table 1 that the auto-covariance of the water consumption with the temperature and precipitation reduce significant after lag two. It is therefore reasonable to believe that the capability of linear model to forecast water consumption two days ahead will break down.

Table 1. Auto-Covariance of Water Consumption with Temperature and Precipitation

	Lag 1	Lag 2	Lag 3
Water Consumption with Temperature	0.82	0.60	0.42
Water Consumption with Precipitation	0.82	0.60	0.42

The lag measures the correlation of variables over a defined time span. For example, lag-2 measures the correlation of the current water consumption with temperature or precipitation 2 days ago. Since our goal is to forecast water consumption 5 days ahead, the best weather and consumption data we can get are previous 5 days and before. For example, if we wish to forecast water consumption 5 days ahead, say next Friday, then we will use this Sunday's weather and consumption data as input. The input and output variables in this study are listed in Table 2.

Table 2. Input and Output Variables for Water Consumption Forecasting Model

MGD_i	Water consumption i days ago
T_i	Maximum temperature i days ago
$Precip_i$	Rainfall i days ago
Day	Indicate variable, which is 1 for weekend 0 otherwise
Month	Indicate variable with Jan = 1, Feb = 2, …, Dec = 12
i = 0, 1, …, 5	

A four layer neural network was trained using 1994 to 1999 data and tested on 2000 data. The results are compared with the linear model. Table 3 lists the performance of linear and neural network models on testing data.

Table 3. The Performance of Linear and Neural Network Model on 2000 Testing Data

Lag	R^2 of Linear Model	R^2 of Neural Network
1	0.70	0.72
2	0.23	0.41
3	0.08	0.28
4	0.02	0.17
5	0.001	0.15

It can be seen that the neural network model outperforms the linear model slightly for lag 1, but significantly outperforms the linear model after lag 2 on testing data. This indicates that the forecast of water consumption one-day ahead using the neural network model is almost as good as the linear model, but it outperforms the linear model significant for two to five days ahead.

Discussion and Conclusion

In this paper I demonstrated that neural network model is a useful tool for estimating missing hydrological data and for forecasting short-term water consumption. In both cases, the neural network model outperformed linear model. In the first application the neural network model successfully estimated extreme peak flow, which is particularly important during the summer months. In the second application the neural network model can adequately forecast water consumption two days ahead. Though the R^2 for lag 2 is only 0.41, the errors were within +15% and –15% range for entire July, 2000. Compared to the 36% error of linear model, neural network's performance is reasonable. After the lag 2, the accuracy of forecasting collapsed for both linear and neural network models. There are certainly some rooms for improvement of forecasting model. One way to improve the forecasting accuracy is to use the forecast of day one as the input to forecast day two, and repeat the process to day 5. But this involves to use forecasted weather data, which is also inaccurate during the summer months. However in practice during 2000 summer season the neural network model appears to be robust enough to hold the inaccuracy of weather data except in the extreme case. Another way to improve the forecasting is to include cloud covering, wind speed and minimum temperature in the models. It remains to see how those additional input variables will affect he forecasting accuracy.

Neural network model holds several advantages over other statistical models in data analysis. First, neural network model processes data purely based on the information presented in the relationship between input and output without pre-assumptions, such as normality or other statistical distributions of the data. Second, neural network model possesses the ability of learning from the historical behavior of the data. This feature enables neural network model being more robust in estimation and forecasting data that has not been seen yet. Finally, the neural network is more flexible in terms of architecture and learning processes.

References

Baxter, C. W. et al. (2001). "Drinking water quality and treatment: the use of artificial neural networks." Can. J. Civ. Eng. 28(Suppl. 1): 26-35.

Brion, G. M., T. R. Neelakantan and S. Lingireddy (2001). "Using neural networks to predict peak Cryptosporidium concentrations." Jour. AWWA, 99-105.

Garrett, J.H., D. J. Gunaratman and N. Ivezic (1997). "Introduction in artificial neural networks for civil engineers: Fundamentals and Applications." Edited by N. Kartam, I. Flood and J. H. Garret. New York, NY: ASCE

Hagan, M. T., H. B. Demuth, and M. Beale (1996). "Neural network design". PWS Publishing Company, Boston, MA.

Schwerk, T. (1996). "Forecasting of time series with neural networks." Z. Angew. Math. Mech 76: 219-222.

Skeel, T., et al. (1998). "Water conservation potential assessment". Seattle Public Utility, Seattle, Washington.

Yabunaka, K., M. Hosomi and A. Murakami (1997). "Novel application of a back-propagation artificial neural network model formulated to predict algae bloom." Wat. Sci. and Tech. 36(5): 89-97

Integrated Watershed Management in Chi-Tou Forest Ecological Area

Shi-Chou Chen[1], Ya-Nan Wang[2] and Chyi-Rong Chiou[3]

Abstract

Chi-Tou Forest Recreation Area (FRA) was one of the best forest recreation areas in the past. Due to the natural disaster, we need to find out the new direction of Chi-Tou for future. The purpose of this paper was to point the guidelines for the future development consideration under the concept of integrated watershed management. Firstly, the concept of ecology should be considered in the engineering constructure. Next, in order to maintain the healthy forest, the concept of ecosystem management should be implemented. Thirdly, natural educational functions have be enhanced. Finally, combine the local community to develop the ecotourism program. These guidelines are under.

Introduction

Chi-Tou Forest Recreation Area (FRA), located in central Taiwan and part of the experimental forest, National Taiwan University, is the first forest recreation area in Taiwan. This recreation area was established in 1970, and has been developed for many years, so each of the recreation facilities and public facilities is constantly improved. Since 1977, the number of annual visitors has averaged around one million, making this one of the country's most popular forest recreation area. However, it always has the potential problem when the development is over. Chi-Tou FRA was forced to close to public twice in the last three years. One was due to 921 earthquake in 1999, the other one was due to Toraji typhoon in 2001. Until now Chi-Tou FRA is closed to public. However, Chi-Tou FRA has close relationship with local community in the past two decades. When Chi-Tou FRA was closed the local economical condition was drop very quickly. Therefore, the pressure of local community is too strong to wait for perfectly recovery. Now, the time was set to re-open on 12/01/2001. However, Chi-Tou FRA is a very good case for watershed management issue. We need to plan the future Chi-Tou FRA based on the viewpoint of watershed management.

---

1. Director of Forestry Department, Council of Agriculture, Taiwan
2. Professor, Department of Forestry, National Taiwan University
General Director of the Experimental Forest, National Taiwan University
3. Assistant Professor, Department of Forestry, National Taiwan University

Toraji Hazards

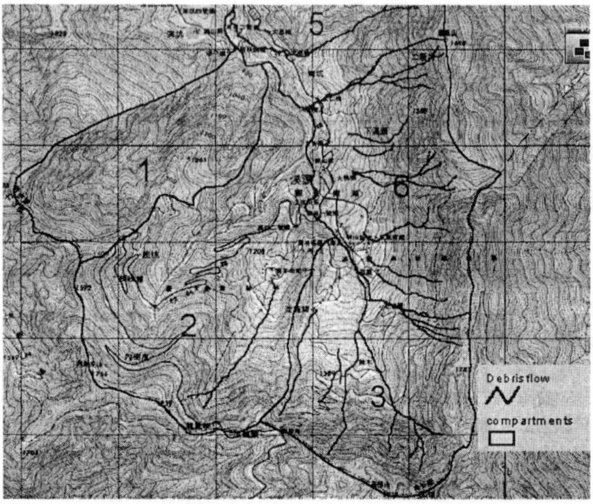

Fig. 1 The disaster map of Chi-Tou FRA

Toraji typhoon visited Chi-Tou FRA on 7/30/2001 and caused very serious disaster. Based on the Cheng's report (Cheng *et al.*, 2001), there were 17 debris flows in Chi-Tou areas. The landslide areas are 135.8 ha in the 1482.9 ha of watershed. Therefore, Chi-Tou FRA was closed until now. The recovery cost was estimated more than US$30,000,000.

The potential problems in management of Chi-Tou FRA

Even most of reports indicated Toraji disaster is continuous from 921 earthquake in 1991, we have some evidences to support that the over development might be the one of the potential reasons to enlarge Toraji disaster. In order to fit the recreation demand of public, Chi-Tou FRA developed rapidly in the past thirty years. Of course, many problems were hidden under the false face of development. The following issues were thought the potential serious problems in the development of Chi-Tou FRA:

(1).the development of Chi-Tou was over the ecological loading
(2).the recreation quality was decreasing
(3).the relationship with the local community was very bad

The future of Chi-Tou

Chi-Tou FRA was so important to local community that the pressure from local community asked the central government to face the problem of recovery of Chi-Tou FRA. The central government has promised that the budget for recovery will be set up for the next several years. In the future, we will base on the concept of integrated watershed management to review the activities in Chi-Tou area. The following issues should be followed as guidelines for developing the action plans:

(1). Based on the view point of ecology to restore the environment
The restoration construction will be preceded in the next 5 years. In the planning stage, the design of restoration construction will base on the viewpoint of the whole watershed. The ecological engineering should be put on the top priority in the design stage. However, the balance between engineering and nature should be expected in the future. Hopefully, the Chi-Tou area will be a demonstration site of ecological engineering.

(2). Based on the concept of ecosystem management to maintain the healthy forest
Forest is the base of Chi-Tou area. The over-mature plantations is getting more and more. How to maintain the healthy forest will be the big challenge in the future. Ecosystem management is a good way to solve the problem. Thus, implementation of ecosystem management is not an easy thing to do. We need to learn more to maintain the healthy forest based on the concept of ecosystem management.

(3). Based on the concept of environmental education to enhance the natural educational function
Chi-Tou is part of the experimental forest, NTU(NTUPF). The four major objectives of NTUPF: academic research, teaching and field practice, environmental conservation and demonstration in forest management. The purpose of teaching and practice is very important to students and publics. Additionally, the environmental education to public is getting important now. The different natural educational programs will develop to enhance the natural educational functions. These programs will be divided into several classes for the different levels of target student.

(4). Based on the local community to develop the ecotourism program
The ecotourism is the new direction of the traditional forest recreation area. Therefore, we need to combine with local community to develop the different ecotourism programs. The experience of tourist will be fresh. Additionally, through the good interaction between management agency and local community, the relationship could be improved.

Conclusion

Chi-Tou FRA is standing at crossroad now. There are so many ways for the future direction of Chi-Tou. How do we select? We need to have more discusses about it. In basic, the concept of integrated watershed management is no one to against. How to implement it? We need to develop a series of action plan to carry out the concept of integrated watershed management. We are not only from the point of engineering but also from the point ecology and society. Therefore, we point out the four principles as the guidelines for the development of the action plans.

Of course, we need to learn more to balance the nature and human. Therefore, the adaptive management will be sure that the direction is on the right track.

Reference

Cheng et al., 2001, Damage Investigation of Debris in Chi-Tou. (In Chinese)

Heathcote, I.W., 1998, Integrated Watershed Management : principles and practice, John Wiley & Sons, Inc.

Loomis B. J., 1993, Integrated Public Lands Management : Principles and Applications to National Forests, Parks, Wildlife Refuges, and BLM lands, Columbia University Press.

Low Impact Development (LID) Technology For Ultra Urban Areas

Michael Clar*,P.E.

Abstract

Low Impact Development (LID) is an innovative technology, which combines a hydrologically functional site design with pollution prevention measures to compensate for land development impacts on receiving waters. LID is being demonstrated to be very effective for new and urbanizing areas. Some of the fundamental principles of LID technology, specifically the source control approach and the use of micro-scale control techniques have the potential to provide substantial benefits for the stormwater quality retrofits in existing ultra urban areas.

This paper summarizes the results of a number of demonstration projects, and introduces a strategic framework for the potential application of LID technology to ultra urban areas. It describes the hydrologic characteristics of ultra urban areas and relates these to identified stormwater management goals and objectives in ultra urban areas. Also potential LID strategies and techniques, which can be applied in these areas, are identified and described.

What is LID?

Low Impact Development (LID) is a new approach to the control of impacts of stormwater (P.G. Co., 1997, EPA 2000). It is based on the use of micro scale stormwater management practices that are integrated throughout the site to manage the volume, frequency, and quality of stormwater runoff. The goal is to replicate the predevelopment site functions of storage, infiltration, and timing by using these distributed controls that are installed on each lot, or at the source. The micro scale controls that are used in this approach can take virtually any form or be "customized" to address almost any stormwater quantity or quality control problem.

This use of "customized" source controls is a powerful concept. Now almost any feature in the urban landscape (e.g., buildings, roads, walks, open space) can be modified to stormwater control objectives (e.g., recharge, filtering, volume control, peak runoff controls.

---

*President, Ecosite, Inc., 3222 Old Fence Road, Ellicott City, MD 21042
PH 410-804-8000; mclar@smart.net

Why LID In Ultra Urban Areas?

The development of the LID concept for urbanizing areas included a number of site specific small-scale demonstration projects, which suggested that LID technology could be applied to ultra urban areas. Some of these demonstrations included the following:

- Control of a commercial site parking lot expansion at Beltway Plaza (Davis, et al., 1998)
- Retrofit of office building parking lot, Inglewood Center(Minami & Davis, 1999)
- Control of Institutional site parking lot, Tampa Bay Aquarium (Rushton, 1999)
- Retrofit of streetscape components, Bladensburg Port Towns revitalization
- Retrofit of streetscape components for street surface storage, Skokie and Willamette, Illinois. (Walesh and Carr, 1999).

Case Study Summaries

These demonstration projects, a number of which were retrofit projects suggested that a strategy for applying LID concepts to ultra urban areas was worthy of consideration. Some of the major findings of these case studies are summarized below.

Study No. 1: Control of Commercial Parking Lot Expansion at Beltway Plaza.

The Beltway Plaza Expansion project demonstrated the application of bioretention techniques for an infill development project. Infill development projects are very commonplace in existing urban areas. Often these sites have an existing storm drainage infrastructure designed to convey the 2 or 10 year storms, but typically lack either storm water management or water quality control. This was the case at the Beltway Plaza located in Prince George's County, Maryland. It was determined that this site would receive a waiver for peak discharge control due to the existing storm drainage system, but would require water quality control.

The entire parking lot for the expansion received water quality control through the use of bioretention cells. This site proved to be a significant milestone for the use of LID technology for a number of reasons. First it demonstrated that the bioretention concept could be used to control a large parking lot area by breaking the area up into a number of small drainage catchments, each approximately ¼ to ½ acre in size; the micro-scale management concept. Second it has provided crucial information on the durability of this practice. The bioretention cells have been in operation since 1993 and are holding up very well. Third, it provided a site that could be monitored to document the pollutant removal performance of bioretention cells as described below. Fourth, it has provided actual data that verifies the very low and simple maintenance requirements and costs of these practices which are approximately $200.00 per site per year.

The pollutant removal performance of the bioretention cells at the Beltway Plaza location were field monitored and as shown in Table 1, verified that they not only met the water quality control criteria, but actually ranked as the most effective pollutant removal BMP available. In addition the success of this site suggested that bioretention could be an effective retrofit BMP for existing urban areas.

Table 1 Pollutant Removal Performance of Bioretention Practices
(% Removal Rates)

	Cu	Pb	Zn	P	TKN	NH4	NO3	TN
Upper Zone	**90**	**93**	**87**	**0**	**37**	**54**	**-97**	**-29**
Middle Zone	**93**	**99**	**98**	**73**	**60**	**86**	**-194**	**0**
Lower Zone	**93**	**99**	**99**	**81**	**68**	**79**	**23**	**43**

Source: Davis et al., 1998

Case Study No. 2: Parking Lot Retrofit with Bioretention

The parking lot of the Prince George's County Office Complex in Largo, Maryland had been designed with conventional end-of-pipe SWM controls. The Prince George's County government conducted a bioretention retrofit project to demonstrate that the bioretention cell was an effective way to retrofit existing urban areas to provide water quality benefits. This demonstration site also served as an important milestone in the development of LID technology in Maryland for a number of reasons.

First it demonstrated that existing impervious areas could be cost effectively retrofitted with bioretention cells to provide water quality control. Second it demonstrated and documented the use of various materials for the construction of the cells. Third it provided another site that was monitored to document the pollutant removal performance of bioretention cells in a retrofit setting.

Case Study No. 3: LID for Commercial Applications – Florida Aquarium

The Florida Aquarium site is an 11.5 acre, asphalt an concrete parking area that serves approximately 700,000 visitors per year. An innovative stormwater management system was designed using a SWM treatment train approach that incorporates many elements of LID design (Rushton, 1999). Runoff is controlled using the following BMPs.:

1. End-of- island bioretention cells
2. Bioretetion swales located around the parking perimeter
3. Permeable paving
4. Bioretention strips between parking stalls

This project has incorporated a long term monitoring program that is providing valuable performance data on the hydrologic and water quality performance of this facility.

A total of 16 storm events were monitored at the Florida Aquarium site during 1998. The Southwest Florida Water Management District measured rainfall and flow from each of the subcatchments in the parking area and collected ware quality samples on a flow-weighted basis. Comparisons between pavement areas controlled by the BMPs and uncontrolled asphalt areas were made for peak runoff rate, runoff volume, runoff coefficients, and water quality. Sediment cores from swales also were collected and analyzed.

The parking areas controlled by the BMPs showed a significant reduction in runoff volume and peak runoff rate. Table 2 shows pollutant reductions for three pavement types; reduction is compared to pollutant concentrations in runoff from a basin without a swale.

Table 2. Load Efficiency of Pollutants Expressed as Percent Reduction for Three Types of Pavement at the Florida Aquarium Site

Constituents	**Percent Pollutant Reduction***		
	Asphalt w/ swale	Cement w/ swale	Permeable w/ swale
Ammonia	45	73	85
Nitrate	44	41	66
Total Nitrogen	9	9	42
Ortho-phosphorus	-180	-180	-74
Total Phosphorus	-94	-62	3
Suspended solids	46	78	91
Copper	23	72	8
Iron	52	84	92
Lead	59	78	85
Manganese	40	68	92
Zinc	46	62	75

* The basins with swales were compared to a basin without a swale to determine the amount of reduction in pollutant loads possible using these small alterations. Notice that the efficiencies or phosphorus are negative, indicating an increase in phosphorus load in the basins with a swale. This emphasized the importance of source control as well as runoff volume reduction.

Ultra Urban Issues

Existing vs Proposed Development. Ultra urban areas are older established areas. Stormwater management technology and BMPs have typically addressed only new development and thus tend to have a predominantly suburban perspective. To date, with few exceptions, such as the City of Alexandria's BMP Supplement to the Northern Virginia Stormwater Management Manual, there has been very little guidance available for existing urban areas (Alexandria, 1992). Development in existing ultra urban areas often consist of redevelopment of existing site or infill of small parcels.

Most of the traditional BMP technology is focused on the use of ponds for peak discharge and water quality control. These ponds tend to require relatively large area requirements, which represent a conflict in ultra urban areas. Consequently a need clearly exists to reexamine and identify available and appropriate BMP technology for existing ultra urban areas.

Heat Island Effect. A unique issue for ultra urban areas is the growing awareness of and concern related to the heat island effect. It is becoming recognized that the heat island effect creates undesirable public health conditions, related to heat stress, for residents of ultra urban areas. Some of these impacts can be addressed by use of certain green building technology such as roof meadows, which can serve dual purpose by also providing management of urban runoff.

NPDES Phase 1 Permit Requirements. Most major American Cities and adjoining jurisdictions fall under the regulatory requirements of the EPA NPDES Phase 1 permitting process. This process which is currently entering the second permit term (5 years) is requiring permit holders to develop prioritization of water quality concerns at a watershed level and the development of restoration strategies for these watersheds. Currently available restoration strategies for existing urban areas tend to be traditional end of pipe control strategies that require the use of SWM ponds. These ponds are often difficult to site in existing urban areas, and they may often require the removal or destruction of existing wooded or wetlands areas. Consequently small-scale micro management retrofit appears to offer an alternative approach to watershed restoration, which needs to be evaluated.

CSO vs SSO. Most existing urban areas were developed using combined storm and sanitary sewer systems (CSO's), separate storm and sanitary sewers systems (SSO's), or a combination of these two approaches. It should be recognized that the presence of CSOs or SSOs present somewhat different technical challenges and watershed planning objectives. One of the major differences between these two systems stems from the fact that older SSO areas only provide conveyance of storm drainage systems but provide no water quality treatment. Consequently for these areas, water quality retrofitting will become one of the primary objectives for NPDES watershed retrofitting strategies.

In contrast areas developed with CSO's typically provide water quality treatment for small runoff events, i.e., first flush. However these combined treatment systems are rapidly overwhelmed during substantial storm events, often as small as one inch or more of rainfall, which leads to CSO discharges on untreated runoff to receiving streams. Consequently for these areas, retrofitting for volume control of small to medium sized rainfall events ranging from ½ inch to 2 or 3 inches of rainfall will become one of the primary objectives for NPDES watershed retrofitting strategies.

Basic Analysis Unit – The City Block

A typical City Block, or block level, can be considered the most appropriate scale to initially assess the feasibility and cumulative effects of LID for ultra urban areas. The use of the block unit allows for planning for control strategies at a small scale so that incremental and targeted management strategies at the catchment or sub-basin scale can be developed. This is different than a conventional approach that may aggregate many distinct land use patterns into a homogeneous watershed description of functional unit. Each block can be divided into management/ownership areas of lots and public space. The components of each lot includes; buildings, parking areas, landscapes areas, and walks. Public space can be divided into parking areas, sidewalks, streets and open space. Once this is done, then specific engineering and BMP strategies can be tested on each component of the system. When the basic strategies are developed, the assumptions can then be tested throughout the study area.

Ultra urban city blocks are characterized by predominantly high-density land uses with corresponding high levels of impervious surfaces. Identifying BMPs that are appropriate for the major components of a typical city block can facilitate selection and evaluation of appropriate BMPs. These components include; 1) rooftops, 2) parking lots, 3) streets, sidewalks and alleys, and 4) vegetated pervious areas such as, grass strips, yards, and tree planting areas. A brief review of BMPs that are suitable for these individual city block components is provided below.

Rooftop Runoff Management BMPs

Rooftop runoff management BMPs refers to a series of techniques, which involve modifications to conventional building design, that retard, and in some cases treat runoff originating from roofs (Miller, 1998). The modifications include vegetated roof covers, roof gardens and meadows, and roof ponding areas. Roofs are one of the most important sources of concentrated runoff from developed sites. If runoff is retarded at the source, the size of other BMPs throughout the site can generally be significantly reduced in size. Rooftop runoff management effectively increases the time of concentration of runoff derived from roofs, delaying runoff peaks and lowering runoff discharge rates. In highly urbanized areas, rooftop measures may be the only practical alternative for relieving pressure on overtaxed storm sewer systems.

Managing rooftop runoff is of the greatest benefit in highly urbanized settings where space for other BMPs is limited. In addition to achieving specific stormwater runoff management objectives, rooftop runoff management is also aesthetically and socially beneficial. Rooftop runoff management measures are suitable for flat or gently sloping roofs. Furthermore, rooftop runoff management techniques can be retrofitted to most conventionally constructed buildings. Rooftop management, particularly roof gardens, can be effectively used to address the heat island effect of large metropolitan centers.

Rooftop Runoff Treatment and Disposal. The rooftop management BMPs described above are focused primarily on controlling the volume and peak discharge of runoff. These rooftop management BMPs can be coupled with a series of additional water quality control measures to from a BMP treatment train which collectively addresses both the quantity and quality of urban runoff . These additional water quality BMPs include; 1) dripline planters, 2) bioretention cells, 3) dry wells, 4) filters such as the perimeter and underground filter.

Dripline Planters. Dripline planters refers to a number of landscaping features that can be placed to receive runoff from roofs to water vegetation. Possible configurations include foundation plantings below the edge of the roof or above ground box planters along the side of a building. Runoff can be conveyed to the planter from downspouts or directly as a spray using rain dispersers instead of gutters. Additional storage can be obtained placing a coarse medium at the bottom of the box planter. The volume of the planter should be estimated according to the area of the roof. If using foundation plantings, care must be exercised to avoid leakage into basements.

Bioretention Cells. Bioretention cells are small-scale multi-functional landscaped shallow depressions that incorporate a sandy soil mixture with a high infiltration rate. (ETA, 1993; P.G Co., 1997; EPA, 2000; MDE, 2000). They can be used to control runoff volume, runoff timing, and can remove pollutants through the physical, chemical and biological processes that occur in plants, soil, and mulch.

Dry Wells. A dry well consists of a small excavated pit backfilled with aggregate, usually pea gravel or stone(EPA, 2000). Dry wells function as infiltration systems used to control runoff from building rooftops. Dry wells provide the majority of treatment by processes related to soil including adsorption, trapping, filtering, and bacterial degradation.

Filters. There are a number of filter BMPs that can be used (MDE, 2000). Two of the more suitable filters for this application are the bioretention cell (see above) and the perimeter sand filter, also called the Delaware Sand Filter. This BMP can be used to filter stormwater, provides some detention, and depending on site conditions also be designed to provide infiltration. The practice consists of two parallel trench-like chambers usually installed along the perimeter of parking lots. The first or pretreatment chamber has a shallow permanent pool of water. When this chamber fills, it spills over into a sand layer with an underdrain that flows into a clear well prior to discharge. Runoff is treated through sedimentation and filtering.

Parking Lot Runoff Management BMPS

Parking lots can take up a considerable amount of space in ultra urban areas and provide very good potential for the use of ultra urban BMPs. Some of these BMPs include: 1) Porous and/or Modular Paving, 2) Bioretention Islands, and 3) Parking Lot Filter (Baltimore, 2001)

Porous and / or Modular Pavement. Porous and/or modular pavement are BMP systems that can provide the same functional characteristics as conventional pavement systems for parking areas, roads and sidewalks while allowing for the storage, treatment and/or infiltration of surface runoff. There are a number of applications of this BMP which include; 1) porous asphalt pavement, 2) pervious concrete paving, 3) concrete grid and modular paving, and 4) parking groves.

Porous asphalt pavement consists of an open graded coarse aggregate cemented together by asphalt cement into a coherent mass, with sufficient interconnected voids to provide a high rate of permeability to water (Thelen, 1978; Rockville, 1982).

Pervious concrete consists of specially formulated mixtures of Portland Cement, uniform open graded coarse aggregate and water. This material may be combined with certain water reducing and retarding or accelerating admixtures along with air entraining agents. When properly handled and installed, pervious concrete has a high percentage of void space, which allows rapid percolation of liquids through the pavement (Florida, 1993).

Concrete grid and modular pavement is a pavement consisting of strong structural materials having regularly interspersed void areas, which are filled with pervious materials, such as sod, gravel or sand. These applications can be used to reduce water pollution from low-volume traffic areas by providing a bearing surface having adequate strength to accommodate vehicles while allowing infiltration of surface water and filtration of pollutants.

Parking Lot Storage. This is a BMP that can be used to provide temporary surface storage and controlled release of stormwater runoff on paved (impervious) parking areas or within parking lot landscaped islands. This practice can be used where portions of large, paved parking lots can be temporarily used for stormwater storage without significantly interfering with normal vehicle and pedestrian traffic. Shopping centers and large employee parking areas are likely places for use of this measure. Wherever parking lot landscaped islands are required.

Parking Lot Filter. This control measure is simple an application of the perimeter sand filter also called the Delaware Sand Filter, filters stormwater and provides some detention. The practice consists of two parallel trench-like chambers usually installed along the perimeter of parking lots. The first or pretreatment chamber has a shallow permanent pool of water. When this chamber fills, it spills over into a sand layer with an underdrain that flows into a clear well prior to discharge. Runoff is treated through sedimentation and filtering (Baltimore, 2001).

Streets, Sidewalks and Alleys BMPS

Streets, sidewalks and alleys also make up a very substantial percentage of the typical city block. Historically very little guidance has been provided for managing these components. However there are a number of innovative BMP applications, which can be used to manage runoff from these areas. These BMPs include: 1) Alley Filter, 2)Grated Sidewalk Infiltration System, 3) Pedestal Sidewalks, 4) Inlet / Manhole Filter System.

Alley Filter. The alley filter is a very simple application of a filter such as a perimeter sand filter or an underground sand filter to mange runoff as close as possible to its source and provide treatment prior to entering the storm drainage network. This application is particularly well suited for areas with SSOs, which do not treat urban runoff (Baltimore, 2001).

Grated Infiltration Systems. In some cases, grates are installed to protect trees located along sidewalks or in parking areas and can be retrofitted to provide off-line stormwater management using curb cuts to divert runoff. These locations can be used in combination with a grate over an infiltration or filter system. Grated infiltration or filtration trenches can also be installed over pedestrian crossings and along the periphery of parking lots (P.G. Co., 2001).

Grated infiltration/filter trenches usually receive stormwater inputs through a curb-and-gutter system or as sheet flow. Overflow may be directed to other controls or to landscaped areas. A replaceable sand or cloth filter above the in-situ soils improves long-term performance and eases maintenance. The grate must be capable of bearing the weight of pedestrians and or vehicles. In areas were pedestrian traffic is expected, the grating should be alternated with non-grated surfaces to allow safe passage. The grates should be removable to allow for maintenance.

Grated infiltration /filter trenches are used in areas where space is restricted or when the use of extensive landscaping is not desirable. This type of trenches can be used in new development and for retrofits.

Pedestal Sidewalks. The pedestal sidewalk is a variation of modular paving, which can be used for sidewalks. Storage can be made available under flat surfaces by using spacers to create pedestals that support paving units over the sub grade. Water infiltrates through the interstices in the paving units and fills the space underneath. The water is slowly removed through evaporation. The spacers should be stacked to achieve at least one inch of storage. A sufficient number of pedestals must be provided to assure that the pavers are stable. The storage space must be contained in a rigid frame to prevent the pavers from shifting. The pavers should be inspected annually to determine if removal is required to clean the storage space of sediment. This practice can be used wherever pavers are used. The need for inspection and maintenance must be considered when implementing this BMP (P.G. Co., 2001).

Inlet / Manhole Filter System. This BMP is simply a variation of the alley filter. A perimeter or underground filter can be inserted up gradient of any storm drainage inlet or manhole to intercept street runoff and provide treatment prior to discharge to the storm drainage system. This application is particularly suitable for SSO areas that do not have any water quality management for surface runoff. Street runoff is directly into the filter by providing a curb cut, or curb depression, to divert runoff into the filter facility. The treated runoff discharges from the treatment chamber into the storm drain inlet or manhole (Baltimore, 2001).

Street storage. Street storage uses a combination of roadway and storm drain inlet modifications to change the timing of runoff in a watershed. (Walesh and Carr, 1999). Street storage functions by causing water to temporarily pond water in the roadway and curb and gutter. This is typically accomplished by a combination of modified traffic calming devices that traverse the roadway. The depth of ponding is only a few inches so that it does not impede traffic. Storm drain inlets are also usually modified by the addition of infiltration and/or storage vaults and restricted outlet size to detain runoff. Street storage is best used on low volume roads and streets where residents and property owners are aware of any potential issues associated with ponding water such as traffic safety concerns.

Summary

A number of BMP measures suitable for application to ultra urban areas have been identified. These practices will prove useful to NPDES Phase 1 municipalities in the evaluation and formulation of watershed restoration strategies for ultra urban watersheds.

The BMP measures described provide a strategy for controlling urban runoff from the building out to the streetscape. This provides an opportunity to select and deploy a range of onsite micro-scale technologies, which are easier and less costly to integrate into an ultra urban landscape. This approach provides an opportunity to retrofit ultra urban landscapes on a block-by-block basis and can easily be incorporated into routine urban infrastructure repairs and upgrades. These include sidewalk replacement, street resurfacing, curb and gutter replacement, and storm drainage repairs and upgrades.

In order to facilitate and promote the use of these BMPs local public works agencies are encouraged o undertake small demonstrations of these technologies similar to those referred to earlier in this paper. These demonstrations will help the public works staff get familiar with this micro-scale technology and verify its effectiveness in pollutant removal. In addition public works agencies are encouraged to adopt and incorporate standards and specifications for these practices so that local designers will be able to specify these BMPs without creating unreasonable delays in obtaining the necessary reviews and approvals.

References.

Alexandria, Virginia, 1992. *Alexandria Supplement to the Northern Virginia BMP Handbook*, prepared by Department of Transportation and Environmental Services, Alexandria, Virginia, 1992.

Baltimore, Maryland, 2001, *Baltimore City Supplement to the Maryland 2000 Stormwater Design Manual*, prepared by Engineering Technologies Associates, Inc., for the Baltimore City Department of Public Works, Baltimore, Maryland (In progress).

Davis, A., M. Shokouhian, H, Sharma, and C. Minami, 1998. *Optimization of Bioretention Design for Water Quality and Hydrologic Characteristics*. Report 01-04—31032. Final report to Prince George's County, Maryland.

Engineering Technologies Associates, Inc., (ETA), 1993. *Design Manual for Use of Bioretention in Stormwater Management*, prepared for Prince George's County, Maryland.

Florida, 1993, The Florida Development Manual: *A Guide to Sound Land and Water Management*, prepared by the Stormwater / Nonpoint Source Management section, Florida Department of Environmental Regulation.

Maryland Department of the Environment (MDE), *2000 Maryland Stormwater Design Volumes I & II.*

Miller, C., 1998. *Vegetated Roof Covers: A New Method for Controlling Runoff in Urbanized Areas.* Pennsylvania Stormwater Management Symposium, October 21-22, 1998, Villanova University, Pennsylvania.

Minami, C. and A. Davis, 1999, *Evaluation of Pollution removal Characteristics at Bioretention Facilities at Peppercorn Place*, Project No. 011-4-33173, Environmental Engineering Program, Department of Civil Engineering, University of Maryland, College Park, MD, September, 1999.

Prince George's County, Maryland, 1997, *Low-Impact Development Design Manual,* prepared by Tetra Tech, Inc, Laurel, MD, prepared for Prince George's County, MD

Prince George's Co., 2001, *LID Integrated Management Practices Manual*, (In Progress), Prince Georges County, MD

Rushton, B., 1999. *Low Impact Parking Lot Design Reduces Runoff and Pollutant Loads; Annual Report #1*. Southwest Florida Watershed Management District, Brooksville, Florida.

City of Rockville, MD, 1982 . *Construction Specifications for Porous Pavement,* Department of Public Works, Rockville, MD

U.S. Environmental Protection Agency, 2000, *Low-Impact Development Design Strategies: An Integrated Approach*, prepared by Tetra Tech, Inc., Fairfax, VA

Walesh, S., and R. Carr, 1999, *Street Surface Storage for control of Combined Sewer Surcharge*, in WRPMD '99: Preparing for the 21st century, ed. M.E. Wilson, Proceedings of the 26th Annual Water resources Planning and Management Conference, Tempe, Arizona, June 6-9, 1999

DEVELOPMENT OF A BEST MANAGEMENT PRACTICE (BMP) PLACEMENT STRATEGY AT THE WATERSHED SCALE

Shaw L. Yu[1], M. ASCE, and Jenny Xiaoyue Zhen[2], Student M. ASCE

Abstract

Nonpoint source pollution control requires the implementation of best management practices (BMPs). There have been thousands of structural BMPs built in the United States since the stormwater regulations were enacted in the late 1980's and early 1990's. The number will keep increasing in light of that the NPDES Phase II regulations are to be implemented within a few years. Most structural BMPs have been installed at the local level, or "on-site", and not at the regional or watershed level in a comprehensive manner. In recent years, the USEPA is promoting the "watershed approach" in controlling pollution from various sources in a watershed. One important aspect of the watershed strategy should be the optimal placement of the BMPs so that the combined effects of the BMPs will provide the most cost-effective way of treating stormwater. In this paper, a methodology is developed to assist in the determination of BMPs placement strategies at the watershed scale. The AGricultural Non-Point Source Pollution Model 2001 (AGNPS 2001) developed by USDA, is used for the BMP placement analysis. The relative effectiveness of BMP at three different spatial placement levels, i.e. on-site, sub-regional and regional levels, is compared. BMP simulation models are used in connection with AGNPS 2001 to evaluate BMP performance in the context of an entire watershed. Based on the model simulation results, a BMP placement optimization approach is developed to determine a most cost-effective BMP placement strategy at the watershed scale. A hypothetical case study is described in the paper to illustrate the methodology.

Introduction

Best management practices (BMPs) have long been employed as a standard technology for controlling nonpoint sources of pollution. Structural BMPs such as dry and wet detention

1. Professor of Department of Civil Engineering, University of Virginia, Charlottesville, VA, USA 22903.
2. Ph.D. Candidate of Department of Civil Engineering, University of Virginia, Charlottesville, VA, USA 22903.

ponds are the most commonly used BMP. Since the enactment of the various federal, state and local stormwater management and nonpoint pollution control regulations in late 1980's and early 1990's, there have been tens of thousands of detention ponds built in this country. For example, it was estimated that more than 3,000 detention ponds were in use in the metropolitan Washington, D.C. area in 1989 (Yu et al, 1993). In Virginia, for a small city such as Charlottesville, there are over 100 detention ponds scattered all over the landscape at the present time. The number of BMPs will continue to grow in light of the announcement of the federal NPDES Phase II stormwater regulations. Many small municipalities and construction sites, which were not regulated under the NPDES Phase I program, will now be included.

To this date, most BMPs have been installed at the local, or on-site, stormwater management level, instead of the regional, or watershed level. To effectively and efficiently control nonpoint source pollution, the key question that need to be addressed and resolved is how to implement BMPs in a cost effective and environmentally beneficial way to control the multiple types of pollution from different sources at the watershed scale. To put it more specifically, how the BMPs should be strategically placed in a watershed in order to meet the total maximum daily load (TMDL) requirement for an integrated watershed pollution management? The TMDL program is a water-quality based approach, i.e., a process that provides a link between BMP design and actual water quality improvement. Currently, on-site BMPs are designed based on a "performance" expectation such as "80% total suspended solids removal". What the hundreds of BMPs will do "collectively" to impacting the water quality at the watershed level remains to be determined. The incorporation of a BMP placement strategy at the watershed level should help establish such a link, which is the principle goal of the TMDL process.

Literature information on BMP planning at the watershed scale is relatively scarce. One recent study by Tilley and Brown (*1998)* compared the "treatment area" required of constructed wetlands for treating various pollutants at different spatial watershed scales. Three scales were compared, i.e., neighborhood or on-site (10-100 ha); sub-basin or regional (100-1,000 ha), and basin (>1,000 ha). Tilley and Brown applied their method to urbanized watersheds in the Miami, Florida area of the United States. Their results indicated that in general, the basin scale required the least wetland treatment area as a percentage of the total watershed area, the neighborhood scale required the most, and while the sub-basin scale required an area in between the other two. The study provided some insight into the scaling effect on BMP planning and optimization, even though the different design criteria were used for the three scales and no cost-effectiveness was considered.

In this study, several levels of watershed BMP spatial placement strategies were compared. The same basic design criterion was applied to all levels of BMP design. The cost-effectiveness of the various BMP placement scenarios was examined using the unit pollutant removal cost as the index. A hypothetical case example is presented to illustrate the proposed methodology.

Methodology

A methodology was developed to assist in determination of BMPs placement strategies at the watershed scale. It consists of the following steps:

1. Investigate the watershed-wide nonpoint pollutant loading generation to identify the critical sources that contribute most to the water quality problems for the water body in question.

The development of management alternatives in a detailed analysis should start naturally from an inventory of sources. The ideal way of investigating the nonpoint pollution loading is to establish a comprehensive monitoring system. For most situations, however, it's not possible or practical to install a large number of monitoring systems. An alternative approach of nonpoint pollutant loading assessment is to use watershed stormwater models that have the capability of identifying critical source areas within a watershed.

2. Determine potential BMP placement locations. For each potential BMP site, an optional BMP technology is selected, and properly designed based on the corresponding drainage area conditions.

From an integrated watershed point of view, BMPs can be placed at different spatial levels, such as on-site, sub-regional, and regional placement levels. Figure 1 depicts the three spatial levels of BMP application. Potential sites are selected based on land availability and geographical conditions.

3. Evaluate the performance of BMPs in the context of a watershed, or in other words, to estimate the pollutant reduction effectiveness of each optional BMP placement plans at the watershed scale by using a watershed model, with the BMP simulation routine embedded.

4. Determine a cost-effective BMP placement plan by comparing the simulation results from the previous step along with the cost estimation, and by considering other constrains such as space availability.

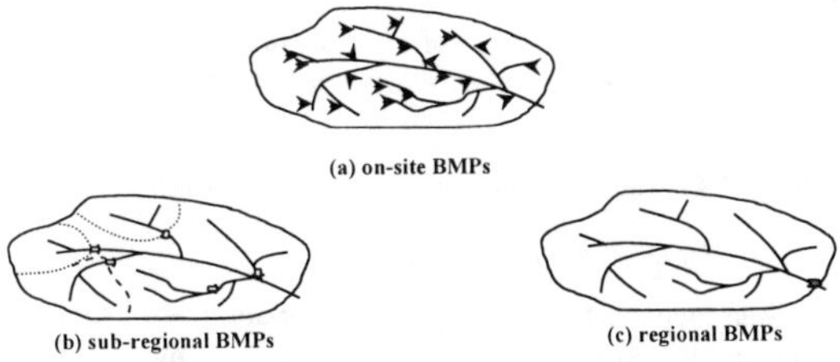

Figure 1. Schematic diagrams showing BMP placement at various spatial levels: (a)on-site; (b)sub-regional; and (c)regional.

Implementation of the Proposed Methodology

Following the steps listed above, an implementation plan was developed to apply the proposed methodology.

1. Employ AGNPS2001 to identify the critical areas that yield the most pollutant loads to the receiving water system.

Many distributed watershed models have been developed to aid in understanding the nonpoint source (NPS) pollution. Among those, AGNPS 2001 (AGricultural Non-Point Source pollutant modeling system, version 2001) is a suite of models developed by USDA (AGNPS 2001, 2001). AnnAGNPS (Annualized AGNPS) pollutant loading model is the key module of the system. It is a batch-process, continuous-simulation computer model operating on a daily time step. The pollutant loading predicted are sediment by particle size class and source of erosion; chemicals-nitrogen, phosphorus, organic carbon, and pesticides. Pollutant loading is generated from land areas (cells or sub-watersheds) and routed through stream systems on a daily basis. Special land use components such as feedlots, gullies, and point sources are included. One distinctive feature of AnnAGNPS is its source accounting function. AnnAGNPS outputs information on total quantities (such as runoff volume, sediment, and nutrients loading) for the entire simulation period at user-selected locations and each location's contribution as a ratio to the quantity at the watershed outlet or other user-specified spot. This information is useful in identifying critical areas, which contribute significant amount of pollutant, sediment or nutrients for example, to the watershed outlet.

Another advantage of AGNPS2001 is its Geographic Information System (GIS) tools that have been developed to generate data needed for AnnAGNPS input files for all cells of a watershed from the spatial data sources, and also provide a good means of visualizing and displaying the simulation results. The GIS tools can automatically estimate most of the required input data. The input data can be derived from three main spatial data sources (files): 1) Digital Elevation Model (DEM), 2) digital soil mapping units maps, and 3) digital land cover/use maps (Bingner and Theurer, 2000; Merkel and Liu, 2001). Use of GIS tools greatly reduces the amount of work, tediousness, and cost of data preparation for implementing AnnAGNPS simulations.

Suspended Sediments/Solids (SS) was selected as the representative pollutant in this paper. Because contaminants often bind or adsorb with fine particulate matter, SS is often used as a surrogate for other pollutants.

2. Determine potential BMP site; select and design selected BMPs.

    - Selection of potential BMP locations

For each critical area identified by AnnAGNPS simulation results, BMPs can be placed potentially at different levels, which can be generally categorized into three levels, namely on-site, sub-regional, and regional. The on-site level BMPs are located right at the points that collect and control stormwater runoff from the critical areas only. The sub-regional BMPs are located further downstream to collect runoff from larger drainage area, which not only include the critical areas, but also other non-critical areas. Moving downstream further along the stream system, the regional BMPs is located to control the runoff from a higher order sub-basin.

Approximately, the drainage area ranges from 10 to 100 ha for on-site BMP, and 100~1,000 ha for the sub-regional BMP, and >1,000 ha for the regional level.

- Design/Sizing of BMP

In this study, for demonstration purposes, dry Extended Detention (ED) basin was selected as the only option of BMP at all placement levels. ED basins are designed to not only reduce the rate of runoff from given design storm events, but also to detain runoff from small, more frequent storm events. In a dry ED pond, a temporary pool forms when runoff from a storm event backs up behind an undersized outlet. The outlet size is selected to yield a given detention time for the design storm, taking into account the pounded volume and average head in the temporary pool.

The dry ED ponds design features were estimated based on the current knowledge of engineering design practices. The pond storage volume was designed to retain runoff from a 10-yr 24-hr storm. An orifice was selected to be the outlet structure and was sized to yield 12-hr detention time for 1-yr, 24-hr design storm. The surface area of the on-site pond was assumed to be 1 percent of the drainage area. For sub-regional and regional dry ED ponds, because they need relatively large surface areas that might not be available at certain locations, two options were used to allow some flexibility in sizing the surface area : 1) the pond surface area is one percent of the whole drainage area; 2) the pond surface area is one percent of the critical area included in the drainage basin. The pond sidewall was assumed to be vertical.

3. Simulate the long-term performance of detention ponds by using the impoundment module in AnnAGNPS.

    - Impoundment module in AnnAGNPS

The impoundment module uses a modified sediment deposition algorithm, which relate the trap efficiency of each sediment particle size classes to fall velocities and detention time, to estimate the amount of sediment removed for each storm event.

    - Impoundment module input parameter preparation

Five input parameters are needed in the impoundment module to reflect pond geometry and outlet structure, which determine the sediment removal efficient of each pond. They are permanent pool depth - h_0, stage-storage power curve coefficient - a , and exponent - b, and stage-discharge coefficient - c , and exponent - d.

For dry pond, permanent pool depth (h_0) is zero.

Stage-storage power curve equation:

$$V = a\,h^{b}$$

Where: V = pool storage, (ha-m); and
h = pool stage, (m).

For vertical wall pond, b equals to 1, and a is the pond surface area.

Stage-discharge equation:

$$Q = c\, h^{d}$$

Where: Q = outflow discharge, m^3/s; and
h = pool stage, (m).

For pressure orifice outflow, d = 0.5. The stage-discharge coefficient (c) can be determined from the total runoff volume (V_I) of the design storm and required detention time (t_d) by:

$$c = 10000\, V_I / [t_d\, (V_I / 2a)^{0.5}]$$

Where: V_I = event runoff water volume, (ha-m);
a = coefficient for the impoundment stage-storage relationship; and
t_d = detention time for runoff event, (s).

- Simulate long-term performance of detention ponds in the context of a watershed

Using AnnAGNPS, various BMP placement scenarios are simulated for a certain time period, for example several years, to evaluate the long-term pollutant removal effectiveness in the context of a watershed. The scenarios can be categorized into five groups: 1) background scenario: no BMPs applied within the watershed; 2) on-site BMP scenario: on-site BMPs are placed to control the critical areas only; 3) sub-regional BMP scenario: regional BMPs are selected as the means to control NPS pollution; 4) combined on-site and sub-regional BMP scenario: both on-site and regional BMPs are applied within one watershed, but to control the critical areas; 5) regional BMP scenario: one regional BMP is placed at the outlet of the regional watershed.

4. Determine an optimal BMP placement plan.

By comparing the pollutant (SS) reduction effectiveness of various BMP placement plans, theirs corresponding cost, and other constraints, an optimal plan can be determined.

- Cost estimation

Detention ponds are relatively inexpensive storm water practices. The total costs, including construction costs and maintenance costs, can usually be related to the pond storage volume. The National Association of Homebuilders has estimated the total annual cost for detention ponds is \$0.28 to \$2.47 per cubic meter storage (\$0.008 to \$0.07 per cubic foot storage).

In this paper, for simplicity and demonstration purpose, it is assumed that the total annual cost for ponds with a storage volume less than feet 28,340 cubic meters (1,000,000 cubic) is \$2.47 per cubic meter storage (\$0.07 per cubic foot storage); and the total annual cost for ponds with a storage volume greater than 28,340 cubic meters (1,000,000 cubic feet) is \$1.76 per cubic meter storage (\$0.05 per cubic foot storage). The total annual cost for each pond is determined by multiplying the pond storage volume and the cost per unit storage volume.

- Cost per unit SS loading reduction

The cost per unit SS loading reduction, was calculated as:

(Total Annual Cost) / (Amount of SS loading reduced annually by applying BMPs)

The optimal BMP placement strategy can be determined by comparing the cost per unit SS loading reduction of each optional plan, under the constrains of total cost and pollutant loading reduction requirement.

Case Study

To further illustrate the methodology for placing stormwater treatment BMPs cost-effectively at the watershed scale, a hypothetical case study is presented.

- Description of study area.

The study area is a 1,172-ha watershed, which has the identical topography as the sub-basin north west of Ivy Creek watershed located in Albemarle County, Virginia. The land slope of this area is very mild. The dominant land uses of the hypothetical watershed are grazed pasture and forest. The study area is covered by one single loamy texture soil type. The statistics of the precipitation data are annual average precipitation = 1,084 mm (42.7 in.); rainfall depth of the 1-yr, 24-hr storm = 56 mm (2.2 in.); rainfall depth of the 10-yr, 24-hr storm = 122 mm (4.8 in.), and rainfall follows SCS type II distribution.

- Critical areas identification.

AnnAGNPS was used to simulate the watershed hydrology and NPS pollutant loading generation, then to identify the critical areas. The study area was delineated into 155 cells, and each cell was assumed to be homogenous in terms of soil type and land cover. AnnAGNPS Input Data Preparation Model (AIDPM) (Darden, R. and R. Bingner, 2001; Darden, R. et al, 2001) was used to calculate the drainage area, average land slope, average elevation, overland flow length and slope, shallow concentrated flow length and slope, concentrated flow length and slope, and length-slope (LS) factor for each cell. The reach length and reach slope were also automatically obtained by applying AIDPM. Three years synthetic climate data was used to drive the watershed simulation. Suspended sediment was selected as the concerned pollutant.

By checking the source accounting output data, 30 cells was identified as critical areas (shown in Figure 2.), which have unit area sediment load ranging from 0.542 to 0.994 ton/ha/yr, which is greater than the watershed average value of 0.379 ton/ha/yr.

- BMP placement, design, and cost estimation.

On-site, sub-regional, and regional BMP sites were selected, as shown in Figures 2. For each site, dry extended detention ponds were determined to be the proper facility for trapping sediment.

Following the dry extended detention ponds design principles presented previously, the pond storage volume was sized to retain runoff from a 10-yr 24-hr storm, and the outlet orifice was designed to yield 12-hr detention time for 1-yr, 24-hr design storm. The pond sidewall was assumed to be vertical. For on-site ponds, the pond surface area was sized to be 1 percent of the drainage area. For sub-regional and regional dry ED ponds, two options of sizing the surface area were applied: 1) the pond surface area is one percent of the whole drainage area; 2) the pond surface area is one percent of the critical area included in the drainage basin. It is obvious that the one percent of the critical area criteria gives smaller pond surface area than that of the one percent of the drainage area.

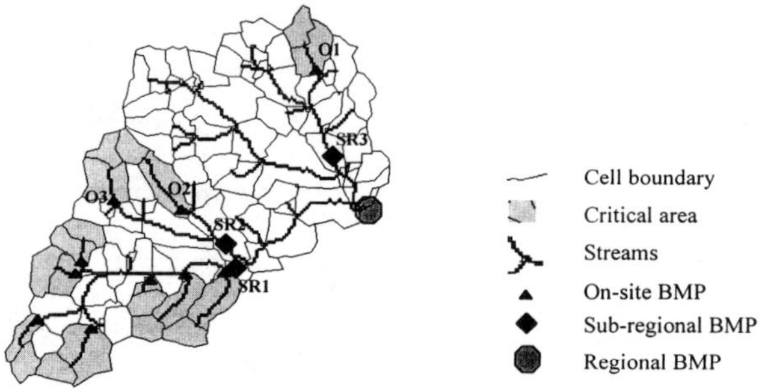

Figure 2. Study area: critical areas and potential BMP placement locations

The total annual cost of each pond was estimated via multiplying the cost per unit storage volume by the pond storage volume. The unit cost per unit storage volume was described in previous section.

- BMP placement scenarios, and evaluation of their sediment removal effectiveness by AnnAGNPS.

Using AnnAGNPS, and its embedded impoundment module, six scenarios were simulated to evaluation the relative sediment removal effectiveness of various pond placement plans. The simulation time period was three years.

Scenario 0: no BMP (background)

Scenario 1: one regional pond;

Scenario 2: three sub-regional ponds;

Scenario 3: two sub-regional ponds (SR1, SR2), and one on-site pond (O1);

Scenario 4: one sub-regional pond (SR1), and three on-site ponds (O1, O2, O3);

Scenario 5: ten on-site ponds were applied.

Simulation results of scenario 0 gave the background SS loading when no BMP was applied. By comparing the SS loading from the watershed of scenario 1 ~5 to scenario 0, the SS loading reduced due to the application of BMPs were determined. The average annual SS loading and SS loading reduction of each BMP placement plan, along with the total annual and unit cost estimation are presented in Table 1, Figure 3 and Figure 4. The unit cost, with the unit of \$/ton/yr, was calculated by dividing the total annual cost by the amount of SS loading reduced.

The data are presented in annual average format, which was obtained by averaging three years' output results.

Table 1: SS loading reduction and cost estimation of ponds application scenarios

Scenario No.	0	1		2		3		4		5
Surface area sizing criteria	--------	1% CA	1% DA	1% CA	1% DA	1% CA	1% DA	1% CA	1% DA	DA=CA
SS loading at outlet (ton/yr)	443.9	139.7	81.4	235.4	205.9	250.8	233.1	275.3	265.6	329.8
SS loading reduced (ton/yr)	--------	304.2	362.5	208.5	238	193.1	210.8	168.6	178.3	114.1
Total annual cost ($)	-------	09,532	09,532	04,440	04,440	92,188	92,188	84,264	84,264	94,220
Unit cost ($/ton/yr)		180	151	324	284	329	301	361	341	502

Note: CA – Critical Area; DA – Drainage Area

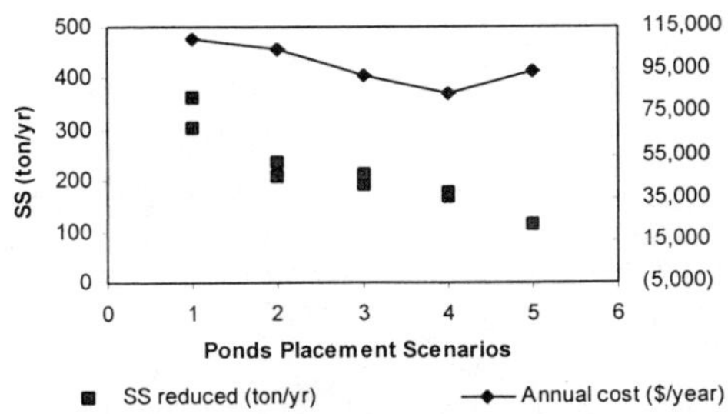

Figure 3. SS loading reduction and annual cost of five BMP placement scenarios

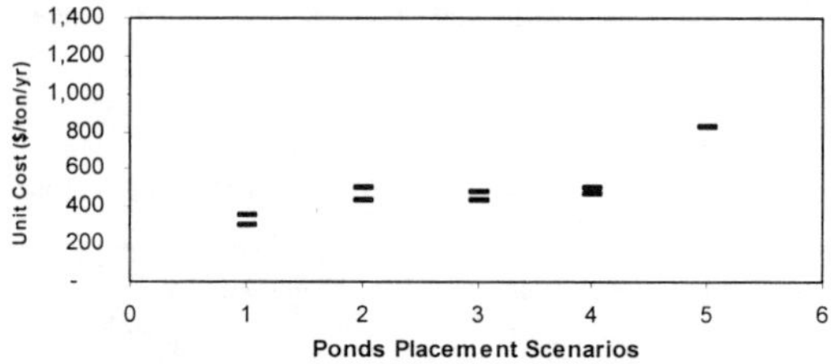

Figure 4. Unit cost ($/ ton SS reduced/yr) of five BMP placement scenarios

Results and Discussion

For each regional and sub-regional pond, two surface areas were tested. From the AnnAGNPS simulation results, it was found that the pond with larger surface area has greater long-term SS removal efficiency, even though the storage volumes and the detention time for the design storm of the ponds were the same. The underlying reason is that the rising height of water pool formed when the pond is receiving stormwater runoff is lower for a pond with larger surface area, thus the detention time would be lengthened because the pond discharge rate is directly proportional to the pool depth. This finding implicated that the surface area of a pond should be sized as large as possible when land is available and other engineering considerations are also satisfied.

Comparing the SS loading reduction and total annual cost of five scenarios (as shown in Figure 3.): The regional ponds can reduce the greatest amount of SS loading from the watershed, but has the highest cost; The on-site ponds reduce the smallest amount, and its total cost is lower than that of regional and sub-regional ponds; Comparing scenario 2, 3 and 4, it is found that along with the number of on-site ponds increases, the amount of total SS loading reduction decreases and the costs also decreases. Among the five scenarios, scenario 4 (one sub-regional pond and three on-site ponds) has the lowest cost, and has greater SS loading reduction than scenario 5 (on-site ponds).

Comparing the unit cost (cost per unit SS loading reduction per year) of each scenario (Figure 4.), it's observed that the regional pond had the lowest unit cost, while the on-site ponds gave the highest unit cost. The unit cost for the other three scenarios were in the middle and tied closely.

As shown by the case study results, regional BMP usually had the greatest pollutant removal capability, but the total cost was also high. Therefore, the application of regional BMP is likely to be limited by the cost constrain, if not by the availability of a large area of continuous land, even though the unit cost of regional pond was the lowest. But if solely consider the unit cost, regional BMP plan was the optimal one. The on-site BMP plan was the last choice that should be considered based on both unit cost and SS loading reduction capability. The sub-regional BMP plan and the combined sub-regional and on-site BMP plans had acceptable unit costs, and their total costs were lower than that of the regional one, thus might be the optimal plan if cost was the limiting constraint.

Conclusions

The purpose of this paper is to present a methodology for identifying a cost-effective BMP placement plan for NPS control at the watershed scale.

The proposed methodology has been shown to be capable of determining the most cost-effective BMP placement strategy under constraints of cost and pollutant loading reduction requirement. It is suggested that the optimal BMP application plan be selected under three considerations, i.e., total cost; pollutant loading reduction requirement, and unit cost. As shown by the case study, of the three BMP placement levels examined, the regional BMP placement was found to be the most cost-effective, while the on-site option the least. However, a combined on-site and sub-regional placement strategy may be the optimal one, considering other constraints, such as total cost and space availability. This shows that the cumulative or synergistic effects of combining BMP spatial placement schemes should be considered in a watershed BMP planning process. It should be noted that which BMP spatial placement scheme is the optimal one may be very case specific, but the methodology should be applicable to all situations.

References

AGNPS2001, 2001, http://www.sedlab.olemiss.edu/AGNPS.html

Tilley, D. R, M. T. Brown, 1998. Wetland networks for stormwater management in subtropical urban watersheds, Ecological Engineering, 10 (1998) 131-158

Yu, S. L, et al, 1993. Stormwater Management for Transportation Facilities, Transportation Research Board, National Research Council, Washington, D. C.

Darden, R., V. Justice, R. L. Bingner, and F. Theurer, 2001. AGNPS2001 Input Data Preparation Model, User's Guide, Version 2., National Sedimentation Laboratory, Agricultural Research Service, USDA

Darden, R., R. L. Bingner, 2001. AGNPS 2001 Input Data Preparation Model, Technical Reference, Version 2., National Sedimentation Laboratory, Agricultural Research Service, USDA

Large Water System Planning and Operation

Kathlie S. Jeng-Bulloch[1], P.E., Jerry Rogers[2], Ph. D., P.E. and William P. Bulloch[3], Jr. P.E.

Abstract

Large urban areas face water supply challenges as they continue to grow. Houston, like many other large urban areas has been in a transition from primary dependence on groundwater to surface water. The transition began in the mid 1970's in order to control land subsidence in Galveston and Harris Counties, where valuable property has been lost.

In order to develop and provide an adequate water supply without total dependence on groundwater, new surface water supplies must be planned implemented. Construction of surface water treatment facilities must be developed, coupled with the expansion of existing surface water treatment facilities and interconnection of transmission and/or distribution mains to ensure safe and reliable distribution systems operation. While groundwater will remain a valuable resource, but pumpage will be limited, primarily to meet peak water demand and provide service to areas where surface water is not available.

The remainder of City of Houston water system consists of groundwater pump stations and associated groundwater wells which originally had minimal interaction with each other.

Safety is also a major consideration in planning and operations of water supply, treatment and distribution system. The impact and response of two major storms, October 1994 and July 2001, will also be reviewed in this paper.

From Surface to Groundwater

The current program of conversion to a water supply system that mainly relies on surface water is reversing the trend of Houston reliance on ground water that began in the late 1880's. In 1906 the

1. Chief Engineer, Dept. of Public Works & Engineering, City of Houston, Houston Texas, U.S.A.
2. Prof., Dept. of Civil Engineering, University of Houston, Houston Texas, U.S.A.
3. Sr. Vice President, Utility Revenue Management, Houston Texas, U.S.A.

City purchased the privately owned Houston Water Works Company system. The City began a program of improvements to the system that included total conversion to well water, metering of water use, and extension of water mains "to keep pace with the City's unusual development". When the City purchased the water supply system, ground water pumpage was increased to about 11 mgd.

The plentiful supply of high quality ground water led to large-scale withdrawals for public supply and rice irrigation in the Houston-Galveston region beginning the late 1800's. After the Houston Ship Channel was opened to ocean-going vessels in 1915, oil refineries requiring large amounts of water were constructed in eastern Harris County, contributing to rapid growth in ground water consumption. By 1940 ground water use had risen to 154 mgd. The use of ground water increased rapidly in the early 1940's as a result of the water needs of major war industries locating in the region. It continued to increase into the early 1950' s as a result of the region's growing industry, population, and rice farming. This growth in ground water use continued until 1954 when surface water conveyance and treatment facilities (Lake Houston, the West Canal, and the first phase of the East Water Purification Plant) were placed in operation.

Symptoms of Groundwater Problems Develop From Growth

Due to the almost total reliance on ground water in the Houston-Galveston region, the symptoms of overuse began to be observed in the 1940's and 1950's. Groundwater quality began to deteriorate. Dramatic drops in the water levels occurred in both major aquifers in the region (the Chicot and Evangline). Land-surface subsidence was beginning to be recognized as a major problem in the low-lying area near Galveston Bay and the Gulf of Mexico.

Ground water pumpage for the region increased from about 154 mgd in 1940 to about 388 mgd in 1954 when major supplies of surface water became available. Ground water usage remained relatively stable until 1962 (398 mgd) and then again increased dramatically to 507 mgd by 1969.

The declines in artesian pressure have resulted in land-surface subsidence of more than one (1) foot in an area of about 3,640 square miles. From 1906 to 1987 subsidence of about nine (9) feet had occurred along the Houston Ship Channel with maximum subsidence of about ten (10) feet in the Pasadena Area in the eastern part of the region.

Surface Water is The Beginning for A Solution

It became obvious to City of Houston water planners that further surface water sources needed to be developed. New facilities included Lake Livingston (completed in 1969 in partnership with the Trinity River Authority) and Lake Conroe (completed in 1973 in partnership with the San Jacinto River Authority).

To utilize the new surface water source being developed at Lake Livingston, the Coastal Water Authority (CWA) was created by a special act of the Texas Legislature in 1967 to develop a conveyance system from the Trinity River to deliver raw surface water to the Houston Area. As a part of the conveyance system for industry and irrigation, the Southern Canal System was purchased in 1969.

By 1976 CWA had completed the raw water conveyance systems which included the Trinity River Pumping Station, the Main Canal, Lynchburg Reservoir, the Lynchburg Pumping Station, the 108-inch Pipeline Ship Channel Crossings, and the Pipeline System to Ship Channel Industries. Surface water was first provided to these industries that year.

Subsidence District Created

The Harris-Galveston Coastal Subsidence District (HGCSD) was created in 1975 by the 64th Legislature to regulate the withdrawal of groundwater within Harris and Galveston Counties. The District was created "for the purpose of ending subsidence which contributes to or precipitates flooding, inundation, or overflow of any area within the district including without limitation rising waters resulting from storms or hurricanes".

In 1976 HGCSD adopted an interim plan designed to have an impact on the subsidence problem in the area most vulnerable to the damaging effects of subsidence. Much progress was made in controlling subsidence in this area known as the ACE, the Southeastern part of Harris County and all of Galveston County, between 1976 and 1984 largely due to the provision of surface water to the Galveston area from the East Water Purification Plant. Water is also provided to the Galveston Area by the Galveston County Water Authority.

Subsidence Regulation from 1985 to 1992

By 1985 the rate of subsidence in southeast Harris and all of Galveston County had dramatically declined as the direct result of reducing groundwater withdrawals. In addition water level increases in the aquifers in amounts up to 150 feet had occurred. However, surface water was not available in the western and northern parts of Harris County. This lack of supply, coupled with extensive growth, had allowed subsidence in those areas to continue at a rather rapid pace.

By 1985 the Subsidence Plan adopted by the HGCSD Board of Directors divided the region into eight regulatory areas, and prescribed the date for each area to convert to surface water use and the percentage of use to be made up of surface water.

In 1989 the Fort Bend Subsidence District was created to regulate groundwater withdrawal in the County which is southwest of Harris County and includes a part of the City. From data available at that time, it appeared that southwest Harris County and northeast Fort Bend County were the primary areas where subsidence would occur based on projected groundwater demand.

More recent data indicates that the growth of water demand in southwest Houston and northeast Fort Bend County is substantially less than previously predicted and the time frame and areas for conversion in Harris County has been revised. In 1992 the HGCSD Plan was revised, which resulted in the number of regulatory areas being reduced to seven and the focus for earlier conversion was shifted from southwest to northern Harris County.

City of Houston Programs

In 1985 the City retained an engineering consulting firm to manage the program for the development of the surface water transmission system. This program was designed to assist the City in meeting the surface water conversion dates established by the HGCSD. From the start of this program through 1997 the City has built or started construction on about 73.5 miles surface water transmission lines ranging from 36-inch to 96-inch diameter, and additional 32 miles of water lines ranging from 12-inch to 30-inch diameter.

The City prepares and updates a Five-Year Capital Improvement Plan for the water system. The City's consultant has also started developing the conceptual plan for surface water conversion for those portions of all subsidence areas that are located within the City. Due to the configuration of the City limit lines in northern and western Harris County the plan, of necessity, is developing a concept for providing surface water in the areas outside the City limits.

In order to supply the required amounts of surface water within Regulatory Areas One through Three, the City has also carried out a program of expansion and new construction of its surface water treatment plants. In 1987 Phase III of the East Water Purification Plant was completed giving it a treatment capacity of about 300 mgd.

In 1990, Phase I of a new Southeast Water Purification Plant with an average day capacity of 50 mgd and a maximum of 80 mgd was completed to provide regional service to several entities in order to meet subsidence criteria for that area. The financing of the construction of this plant was a departure from previous City plant construction. The City solicited participants in a cost-sharing plan for the construction, operation and maintenance of this plant. There are nine (9) participants and the City, with each participant having a different share of the plant capacity. The contracts also provide for the participants to be included for additional capacity that may be constructed. The City currently has a 22% share of the plant capacity.

Because of the multiple capacity ownerships, the participants contracted with an outside firm for the maintenance and operation of the Southeast Plant. The experience gained in developing this plant with outside participation and contract operation will be of great value in the future construction of new water purification plants.

In 1974 the City of Houston determined that another water purification plant located near Lake Houston might be necessary to provide surface water for the northern portions of the City and Harris County. This led to the planning and site acquisition for the Northeast Water Purification Plant.

In conjunction with the plans for the Northeast Water Purification Plant, it became apparent that ultimately additional water, beyond the 130 mgd available in Lake Houston and the City's 60 mgd share in Lake Conroe, would be needed in the San Jacinto River System to supply the needs of the Northeast and East Water Purification Plants. Planning was carried out for a project to divert water from Lake Livingston to Lake Houston taking advantage of the location of Luce Bayou, one of the tributaries to Lake Houston. Again, plans were developed and a Texas Natural Resource Conservation Commission (TNRCC), at that time the Texas Water Commission, permit was obtained. The addition of added capacity to the system for delivering water from the Trinity River to the East Water Plant by completion of the Northwest Lateral in 1991 will allow the East Water Plant to operate, using Trinity River water only, for a significant period of time.

Costs of Conversion

The City of Houston has a forty year history in the conversion from groundwater to surface water use and during that time has assumed the role of the regional supplier. While the achievements have been many, the price has not been cheap. To date the City and its ratepayers have invested, in actual unadjusted construction costs, over $1.8 billion if interest is included.

This has had an immense impact on the City water rate payers. In 1983 the monthly water bill for a customer using 7,000 gallons per month, about the average use, was $11.34. In March of 1993 the same customer was paying $20.47 per month, an 80. % increase in 10 years. Unfortunately, the end is not yet in sight.

Groundwater Today as a Part of the Long Term Strategy

While much emphasis has been and will be placed on the development of the surface water component of the regions water system, groundwater will continue to play a key role in the viability of water service to the region because of its quality and geographic flexibility.

Portions of the City of Houston, adjacent Water Districts and other Cities to the north, west and southwest (of Houston) rely predominantly on groundwater, and will for the foreseeable future. The City of Houston presently operates 192 wells and 97 groundwater plants with a usable capacity of 244 mgd. Even with the present surface water capacity of 328 mgd, groundwater is essential for meeting peak demands and providing for reliable, safe drinking water to the Houston region.

System Integration

Water planners and engineers designing large scale surface water systems to meet the long-term needs of the region. Large diameter transmission mains are being utilized to move surface water long distances to meet Subsidence District conversion criteria.

This process, however, leaves an even larger challenge for the City of Houston and the region. The successful integration of groundwater and surface water in the distribution system requires the modification of existing major groundwater pumping stations and associated reservoirs to accept both ground and surface water.

Water quality and water pressure issues are being dealt with through the implementation of the "Consolidation Plan" developed in 1991 by the City of Houston Water Production staff. This Plan addresses the need to integrate the localized distribution water lines of the original, small, stand alone groundwater systems into a logical set of service areas called "Demand Service Areas"(DSA), compatible with the availability of surface water.

The appropriate blend of ground and surface water in each of these DSA's is being determined based upon water quality and distribution system compatibility. The Water Production staff has developed contiguous demand service areas for the entire City of Houston. Each DSA will consist of 1 major water repump station with associated storage capacity and appropriate groundwater facilities within the DSA. The distribution system for each DSA, which originally consisted of multiple small ground water plants, and localized distribution networks, is now being developed (using network modeling as an analytical tool) following the Consolidation Plan discussed previously.

As small, inefficient and redundant groundwater plants are being removed (based on the Consolidation Plan) to develop each DSA, new water distribution links are being engineered and constructed to maintain or improve water quality and distribution system pressure.

There were 216 wells and 120 groundwater plants in 1985. As a result of the implementation of surface water into the system through the Consolidation Plan and development of DSA's, the are 192 wells and 97 groundwater plants in operation today. These numbers will continue to be reduced as the surface water treatment and transmission main system continues to be developed in the future.

The objective of this operational integration process is to improve the reliability of the water system and water quality through the availability of surface water while striking an appropriate balance between ground and surface water when meeting subsidence goals.

Water Planning

In 1985 a consultant was selected to develop a water master plan that looked forward to the year 2030. The plan concluded that without the Trinity River saltwater barrier a new source of water would be needed in about 2010 and with the barrier about 2030. The existing available source for additional large amounts of water was identified as the Toledo Bend Reservoir on the Sabine River in East Texas.

In May of 1992, the Mayor of Houston convened a Regional Water Futures Assembly that was held in Houston and invited the Mayors of San Antonio and Corpus Christi, the Chairman of the Texas Water Development Board and other interested parties to participate in developing a plan for the transfer and use of surface water in the southeast and south central areas of Texas. This meeting set the stage to begin the Trans Texas Water Program. This program was established to conduct a feasibility study for the transfer and exchange of water and water rights from the Sabine River on the eastern boundary of Texas as far west as Corpus Christi and San Antonio. In February 1993 the City of Houston entered into a joint sponsorship agreement with the Sabine River Authority and the San Jacinto River Authority for the

Southeast Study Area of the Trans Texas Water Project. This study determines the feasibility of the transfer of Sabine River water to the Houston area.

Integrated Resource Planning and Management

We visualize meeting the needs of the future to start by maximizing the use of existing available resources. This requires that we develop a plan that takes advantage of the great groundwater resource available to us. Conversion for subsidence does not require that we turn completely to surface water. The Houston Water Master Plan determined that about 350-mgd of groundwater would still be available for use in the year 2030 when total compliance with the subsidence requirements had been achieved. The Trinity River Salt Water Barrier is currently under construction. When completed it would make available about 260 mgd of raw water by eliminating fresh water releases to prevent saltwater intrusion and by perfecting the water rights assigned to the project. Those water rights are based on a system yield from the combined Lake Livingston Salt-Water Barrier operation.

As we progress in the development of presently planned projects, the system operation of the existing available reservoirs will allow us to make maximum use of available water particularly in periods of prolonged drought. Currently we need to develop a plan for the system operation of Lake Conroe and Lake Houston. With completion of the Saltwater Barrier, a plan for system operation of Lake Livingston and the Barrier will be developed. Finally when the Luce Bayou Diversion is constructed, the three reservoirs and the barrier can be operated as a system.

Conservation was considered as a source of water by the Houston Water Master Plan. The reduction in consumption in effect makes water available for future growth. While conservation is an essential part of the planning, unrealistic reliance on the reduction of water use must be avoided. Water Reuse is another area that will become practical as the cost of providing water increases. We are in fact already "reusing" treatment plant effluent that is discharged into the San Jacinto and Trinity River systems above the reservoirs that serve our water plants. Water utilities should have a great interest in selling the same water twice.

There will need to be significant increases in the capacity of the Southeast and proposed Northeast Water Plants. The output of the existing East Water Plant will need to be maximized and/or expanded. other locations for water purification plants may be identified and utilized. The Houston Water Master Plan included a Southwest Water Plant using Brazos River water. Depending on the location that is selected for delivering water from the Sabine River in the Trans Texas Water Project other water plant locations may be identified and utilized for maximum benefit.

The transmission system for delivery and storage of treated surface water will be vastly expanded to serve the existing developed areas. Further growth in the areas not served by the system will by necessity need to make provision for connection to the system. The Houston Water Master Plan identified a Regional Surface Water Service Area that grew over time to include those areas where groundwater use had reached the maximum allowable. This is a sound concept and though the timing for development of the system has and will continue to change, that growth in system needs consideration in the current planning.

To date our accomplishments have been great, but many challenges remain. As the regional water supplier the past conversions to surface water use has been successful largely because of the comparatively limited number of entities that were participants in the program.

To date no single entity has been identified that can finance a project of this magnitude with such a large number of participants. The method of equitably distributing the cost among the users will have to be developed. This must be done in a manner that does not give an unfair advantage to any of the participants.

Study should also be done to sort out all of the alternatives and select the most practical solution. While the technical design of the system must progress to the point where the magnitude of the cost can be determined, it appears that the real thrust of the study should be directed at the legal, institutional, political and financing issues. These are the issues that need to be resolved before the implementation of a technical plan can be carried out.

The City sees itself as having a role in each of the areas of regulator, supplier and distributor/user. The City is a regulator in the review of plans and the creation of utility districts. The new water conservation plan rules of TNRCC may require that we review individual entities plans for compliance with the rules.

Safety Considerations in the Planning and Operation of Water Systems

The greater Houston area experienced extended intense rainfall and localized flooding from the 15th 0to the 17th of October, 1994. Both the ground and surface water supply and distribution systems were impacted by these occurrences, the groundwater system to a greater extent. Due to flooding of the pump station and associated facilities, service was interrupted to residents in the Belleau Wood subdivision for seven days (from the 15th to the 2nd of October). This subdivision is served by the Belleau Wood groundwater pump station. There was no service interruption to customers who were served by the surface water system.

As a result 9of these flood events, at least 20 people lost their lives, and some 12,000 people were forced out of their homes and into Red Cross shelters and homes of friends and family.

The Belleau Woods subdivision, which is located on the southeast banks of Lake Houston in the San Jacinto River drainage basin, contains 500 plated lots, The City of Houston annexed this property in 1965, and Section I, consisting of 230 lots, was plated in 1966. The Belleau Woods water system consists of an isolated well, closed water supply, and distribution system presently serving 75 customers. Since 1966, the subdivision has flooded on several occasions.

The fundamental problem with the subdivision is that it is located in the floodway. At the time the property was plated, there wasn't the type of information subsequently developed concerning the location of the floodway, floodplain, and the risks inherent in developing and building in these areas. This fact presents continuing operational challenges for the City in providing water service to the subdivision.

The intense rainfall that began on the 15th of October resulted in floodwaters obtaining a height of at least 8 feet above natural ground elevation at the Belleau Woods Pumping Station. The booster pumps, electrical and chemical feed systems were submerged as a result of the flooding. The 5000-gallon pressure tank floated, and pulled way from its foundation and water piping system, which resulted in a seam failure oh the tank.

The water level, however, never rose to the top of the Belleau Woods well seal block, which was set above the height of the 100-year flood.

Besides the fact that the residents of the 75 homes were forced to leave their homes, they were also without water service for seven days (the 16th through the 22nd).

Due to the proximity o f the subdivision to the Lade, the subdivision was flooded for six days, and only accessible be boat. Once the City of Houston Water Production staff had access to the site by land (on the 22nd of October), water service was restored to the subdivision by the 23rd, the next day. The ground water storage tank (56,000 gallons) was half under water during the height of the flooding. As a result, boil water notices were given to the residents that had returned to ensure that there would be no potential health problems while the water system was being flushed and chlorinated.

The cost to bring the system back on line with out the ruptured hydro tank was approximately $90,000. The total cost to restore the Belleau Woods water system to its operating status prior to the flooding with a new 20,000-gallon hydro tank was $107,000.

Water Service Reliability Alternatives

It is clear that due to the location of the Belleau Woods subdivision next to Lake Houston, there will always be potential flooding and water service interruption. These potential service interruption occurrences can be minimized or possible eliminated through the implementation of one of the following alternatives:

1. Remove the water supply, chemical feed systems, storage, and pumping facilities for the 100-year floodplain.
2. Construct facilities inside the subdivision capable of providing water service for a period of up to seven days (minimum) if there is interruption of service for any reason for an extended period of time.
3. Connect the Belleau Woods water system with an alternative supply and distribution system for emergency water service in the event of a service interruption.

Alternative 1.

Remove the water supply, chemical feed systems, storage, and pumping facilities for the 100-year floodplain.

If there is land inside or in close proximity to the subdivision that is outside the floodplain, funds could be allocated to relocate and construct new facilities to ensure that there would be no service interruptions due to flooding events. The estimated cost of this alternative excluding land is approximately $ 200,000.

Alternative 2.

Construct facilities inside the subdivision capable of providing water service for a period of up to seven days (minimum) if there is interruption of service for any reason for an extended period of time.

If an elevated storage tank were to be constructed that would satisfy the basic domestic needs of the residents of the subdivision during an extended event such as a minor flood or major equipment failure, the residents would not be subjected to further potential interruption of service. The estimated cost of an elevated storage tank to serve the residents for up to seven consecutive dates with basic domestic service is approximately $439.000for a tank 150 feet high with a capacity of 100,000 gallons.

Alternative 3.

Connect the Belleau Woods water system with an alternative supply and distribution system for emergency water service in the event of a service interruption.

The Riverside Crest Subdivision adjacent to Belleau Woods has no central water, as each house is served with an individual well. If Riverside Crest or any other subdivision in proximity were to develop a central water system, then Belleau Woods could feasibly develop an interconnect to provide for short-term emergency water service in the event of flooding or other major event. Since this alternative is not feasible at this time, the cost to implement it is unknown.

Relationships of Alternatives to the Safe Drinking Water Act and State of Texas Operating Requirements

The most positive way to ensure that any potential health hazards are minimized is to implement Alternative 1, and move the frailties out of the floodplain. Alternative 1 would be most responsive to the intent of the Safe Drinking Water Act and State of Texas Operating Requirements as a result of this relocation.

Alternative 2 will ensure that there will be continuation of service, both in terms of water availability and quality, but will not satisfy State of Texas operating requirements by still leaving the storage, and pumping facilities inside the floodplain.

If no changes rare made to the existing water system for Belleau Woods subdivision, the residents will continue to experience possible service interruption under certain circumstances such as a flooding event. The existing water supply and pumping system is not consistent with either federal or state regulations.

Development of Kaoping River Basin Management Strategies Based on Calculated Carrying Capacity

Chih-Ming Kao [1], M. ASCE, Sheng-Shou Huang [2], Fang-Chih Wu [2], Ku-Fan Chen [2], Tsair-Fuh Lin [3], E.E. Chang[4], and Pen-Chi Chiang [5]

Abstract

The Kaoping River basin is the largest and the most intensively used river basin in Taiwan. It is 171-km long, drains a catchment of more than 3,250 km^2. Based on the current water quality analysis, the Kaoping River is heavily polluted. Concern about the deteriorating condition of the river led the Government of Taiwan to amend the relevance legislations and strengthen the enforcement of the discharge regulations to effectively manage the river and control the pollution. Investigation results demonstrate that both point and non-point source (NPS) pollutants are now the causes of biochemical oxygen demand (BOD), nutrients, and pathogens in the river. The main water pollution sources are livestock wastewater from hog farms, municipal wastewater, industrial wastewater, NPS pollutants from agricultural areas, and leachate from riverbank landfills. The current daily BOD, NH_3-N, and TP loadings to Kaoping River are 74,700, 39,400, and 5,100 kg, respectively. The calculated daily *Escherichia* Coliform (E. Coli) loadings to Kaoping River are 1.18×10^{16} colony forming unit (CFU). However, the calculated BOD, NH_3-N, and TP carrying capacities are only 27,700, 4,200, and 600 kg per day. The calculated E. Coli carrying capacity is only 4.81×10^{15} CFU per day. Thus, the current BOD, NH_3-N, TP, and E. Coli loadings are almost 2.7, 9.4, 8.5, and 7.3 times higher than the calculated carrying capacities, respectively. To protect public health and improve the river water quality, the comprehensive management and construction strategies are proposed. The proposed strategies include the following measures to meet the calculated river carrying capacity: (1) hog ban in the entire Kaoping River basin, (2) sewer system construction to achieve 30% of connection in the basin within 10 years, (3) removal of 10 riverbank landfills, and (4) enforcement of the industrial wastewater discharge standards. After the implementation of the proposed measures, the water quality should be significantly improved and the BOD and nutrient loadings can be significantly reduced.

Key Words ： water quality; carrying capacity; BOD, hog farm waste, riverbank landfill

1. Associate Prof., Institute of Environmental Engineering, National Sun Yat-Sen University, Kaohsiung, Taiwan
2. Graduate Student, Institute of Environmental Engineering, National Sun Yat-Sen University, Kaohsiung, Taiwan
3. Associate Prof., Dept. of Environmental Engineering, National Cheng-Kung University, Tainan, Taiwan
4. Prof., Dept. of Biochemistry, Taipei Medical University, Taipei, Taiwan
5. Prof., Institute of Environmental Engineering, National Taiwan University, Taipei, Taiwan

Introduction

The Kaoping River basin is the largest and the most intensively used river basin in Taiwan. It is 171-km long, drains a catchment of more than 3,250 km^2, and has a mean flow of 239 m^3/s. The average precipitation is 2,000 mm/yr. Figure 1 shows the Kaoping River, its catchment, and three major reaches (Chi-San Creek, I-Lio Creek, and Lao-Non Creek). It serves as a water supply to the Kaohsiung City (the second largest city in Taiwan), several towns, two counties, and a number of large industries (electronic, steel, petrochemical, etc.). In the meantime, it also receives their treated and untreated wastewater. Taiwan Environmental Protection Administration (TEPA) has developed a three-part classification system (Classes A, B, and C) for Kaoping River based on the purpose of water usage and degree of protection for each stream section (TEPA, 1998a). Table 1 presents the water quality criteria for the three classes in Kaoping River. Basically, the upstream is classified as Class A, mid-stream is Class B, and the downstream near the outfall is Class C. Thus, the highest degree of protection is given to Class A. Recent water quality analysis by EPA indicates that the Kaoping River is heavily polluted (Ning, 1998; EPA, 1999). The concentrations of some major water quality indicators [e.g., biochemical oxygen demand (BOD), suspended solid (SS), ammonia-nitrogen (NH_3-N), total phosphorus (TP), and *Escherichia* Coliform (E. Coli)] are much higher than the Kaoping River water quality criteria. Because of the occurrence of steep slopes in part of the Kaoping River flow courses, higher dissolved oxygen (DO) level is usually observed due to the natural turbulence in the river, which enhances the mass transfer of oxygen between the river and atmosphere (natural reaeration).

Because of the poor raw water quality, the cost for water treatment has been significantly increased by the Taiwan Water Supply Company in the Kaoping region. However, the quality of drinking water is still questioned and complained. Concern about the deteriorating conditions of the Kaoping River led the Government of Taiwan to amend the relevance legislations and strengthen the enforcement of the discharge regulations to effectively manage the river and control the pollution. The Government also requests the Taiwan Water Supply Company to apply advanced water treatment system to provide high quality drinking water to the people in the Kaoping River basin.

The major objectives of this study were to (1) review the recent information on the water usage and assess the water quality, (2) identify the current contributions of point and non-point source pollutants to the river pollution, (3) select an appropriate water quality model for water quality simulation and carrying capacity calculation, and (4) develop river management protocols to improve the river and raw water quality.

Point and Non-point Source Pollution

Results from the review of recent information on the water usage and investigation of the water quality demonstrate that both point and non-point source (NPS) pollutants are now the causes of BOD, nutrients, and pathogens in the river. The main water pollution sources are livestock wastewater from hog farms, municipal wastewater, industrial wastewater, NPS pollutants from agricultural areas, and leachate from riverbank landfills. In the Kaoping River basin, most of the upper catchment is used for agricultural activities including cropland and livestock farming. In the upper catchment, NPS pollutants mainly associated with stormwater runoff from agricultural land uses can be quite diffuse and difficult to treat. Agricultural NPS pollution is considered to be the largest single category resulting in these environmental problems. Nutrients, pesticides, and sediments are the main detrimental NPS constituents. Agricultural NPS pollution is considered to be the largest single category resulting in these environmental problems.

To approach the goals of water resource conservation, water quality management, and land use effectiveness, the regulation of "Drinking Water and Source Water Protection Area (DWSWPA)" was promulgated in 1997. Most of the upper catchment of the Kaoping River Basin has been delineated and identified as the DWSWPA (CPA, 2001). The rule prohibits any developing activities that endanger the quality of source water within DWSWPA. Thus, no agricultural activities are allowed inside the DWSWPA. However, field investigation results reveal that more than 15% of the areas located inside the Source Water Protection Zone have been illegally developed into farm lands. orchard, betel palm farm, and farmland. Betel palm and orchard farms dominate the illegal farmland areas, and are scattered around on both sides of the river corridor (CPA, 2001). Figure 1 shows the photo of the illegal betel palm farm located in the upper reach of Chi-san Creek. NPS pollutants from those farmlands would cause the severe contamination of the source water. Investigation results also show that the NPS pollution contributes more than 10% of the overall pollution loads to Kaoping River (CTC, 1999; TEPA, 2000 and 2001). Thus, NPS pollutants have significant adverse impacts on the river water quality, and should be effectively controlled.

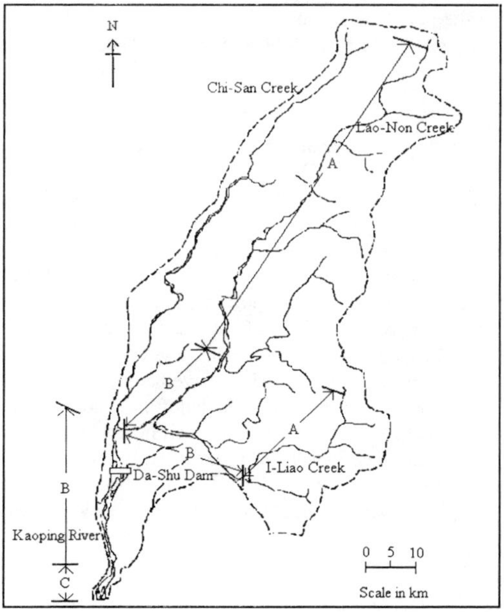

Figure 1. Kaoping River basin and three major reaches.

Table 1. Water quality criteria in Kaoping River.

Water Quality Item	Category A[1] Most upstream reach[1]	Category B Most midstream reach[1]	Category C Downstream reach[1]
DO (mg/L)	>=6.5	>=5.5	>=4.5
BOD (mg/L)	1	2	4
NH_3-N (mg/L)	0.1	0.3	0.3
NO_3-N (mg/L)	10	-[2]	-
Total Phosphorus (mg/L)	0.02	0.05	-
E. Coli (CFU/100 mL)	<50	<5,000	<10,000

[1]Illetrated in Figure 1; [2]No criteria.

Hog farming is a particularly important activity in the rural area of the basin. The estimated hog population in the basin reached 1.7 million in the year of 1996. Most of the untreated hog farm waste was indiscriminately discharged into the river causing the deteriorating of Kaoping River water quality. However, due to the occurrence and overwhelming of foot-and-mouse disease in 1997, the hog population dropped to approximately 0.8 million in 1997. The decrease in hog population also caused the improvement of water quality (TEPA, 1999 and 2000). However, after two years of breeding, the total hog population is estimated to be one million in the whole Kaoping River Basin, and approximately half of the population is in the lower catchment (downstream of the Da-Shu Dam) (CTC, 1999 and 2000; TEPA, 1999 and 2000). Thus, the hog farm waste and agricultural runoff from the cropland have become two major causes of the impairment of Kaoping River water bodies.

Currently, the overall percentage of public sewer system (for collection and disposal of municipal wastewater) connection in Taiwan is only 7%. Due to the rapid urban and industrial expansion, the percent of sewer system connection in the Kaoping River basin is less than 5%. Most of the municipal wastewater in the basin (especially in the rural area) is discharged into the river without treatment. Moreover, there are 229 registered industrial factories discharge their wastewater into the Kaoping River (CTC, 1999 and 2000). However, illegal or expedient discharges are sometimes practiced feeding polluted industrial flows into the river. Therefore, the untreated municipal and industrial wastewaters are also two causes of the poor water quality. In the Kaoping River basin, due to the shortage of available lands for landfill construction, some townships used the riverbank sites as the domestic garbage dumping locations or converted them to simple landfills. Those riverbank landfills (or garbage dump sites) were neither well designed nor well maintained. Thus, landfill leachate could significantly deteriorate the downstream river water quality. There are 10 abandoned riverbank landfills within the Kaoping River floodplain. Leachates from those 10 landfills are currently polluting Kaoping River and might also cause the clogging problems during the flood season.

Water Quality Modeling

In this study, the Enhanced Stream Water Quality Model (QUAL2E) was selected as a water quality-planning tool to perform the water quality evaluation and carrying capacity calculation (Brown and Barnwell, 1987). The QUAL2E Windows interface was developed by U.S. Environmental Protection Agency to assist the implementation of the Total Maximum Daily Load (TMDL) program. It can simulate up to 15 water quality constituents including BOD, nutrients, DO, temperature, algae as chlorophyll A, and coliforms. It uses a finite-difference solution of the advective-dispersive mass transport and reaction equations. A stream reach is divided into a number of computational elements, and for each computational element, a hydrologic balance in terms of stream flow, a heat balance in terms of temperature, and a material balance in terms of concentration are written. Mass is gained or lost from the computational element by transport processes, wastewater discharges, and withdrawals. Mass can also be gained or lost by internal processes such as release of mass from benthal sources or biological transformations. Hydraulically, QUAL2E is limited to the simulation of time periods during which both the stream flow in river basins and input waste loads are essentially constant. QUAL2E can operate as either a steady-state or a quasi-dynamic model, making it a very helpful water quality planning tool. When operated as a steady-state model, it can be used to study the impact of waste loads (magnitude, quality, and location) on instream water quality.

Previous studies have been performed to obtain the input data for the QUAL2E model construction (Ning *et al.*, 1998; EPA, 1999; CTC, 1999). The input data include stream segmentation, locations of inflow and outflow, geological and meteorological conditions, hydrological parameters, decay rates, water quality parameters, dispersion coefficient, reaeration coefficient, BOD removal rate, and benthal oxygen demand (Ning et al., 1998; Chiang et al., 2000). The river was divided into 9 reaches and 85 element (2 km each) to effectively represent the flow conditions. After the model construction, the recent water quality data were used for model calibration (Ning *et al.*, 1998; Chiang *et al.*, 2000). The input parameters for Kaoping River modeling using QUAL2E model are described in Chiang *et al.* (2000). Figures 2a to 2c presents the measured and simulated water quality results for DO, BOD, NH_3-N, and TP in the Kaoping River from the mouth (0 km) to the 84 km upstream location. Results demonstrate that the simulated data had a good match with the analytical water quality results. Based on the investigation, the current daily BOD, NH_3-N, and TP loadings to Kaoping River are 74,700, 39,400, and 5,100 kg, respectively. Results indicate that the hog farm wastes generated from one million hogs contribute 37,900 kg BOD per day, which is more than half of the daily BOD loads to the river. The untreated municipal wastewater also contributes more than 25% of the daily BOD loading. Significant effects of the riverbank landfills on water quality were also observed. Results reveal that landfill leachate contributes 5,000 kg per day of the BOD loading (close to 7% of the daily loads) into the river. Because E. Coli appears to be the most representative of fecal contamination in the water contamination, the colony forming unit (CFU) of E. Coli is also used as the indicator of water quality in Taiwan (Table 1). Figure 3 presents the measured and simulated E. Coli concentrations (CFU/100 mL) in the Kaoping River from the mouth (0 km) to the 84 km upstream location. Results also show that the simulated data had a good match with the analytical results. The calculated current daily E. Coli loadings to Kaoping River are 1.18×10^{16} CFU. Results indicate that the most possible causes of the high E. Coli measurements at some locations come from the hog farm wastes and untreated domestic wastewater discharges.

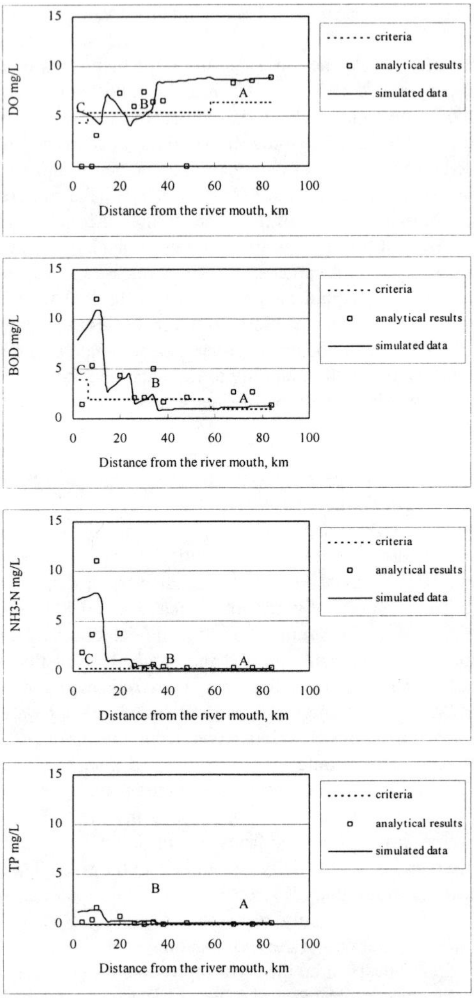

Figure 2. Measured and simulated water quality results (DO, BOD, NH_3-N, and TP) and three water quality classes (Classes A, B, and C) from the outfall (0 km) to the 84 km upstream location.

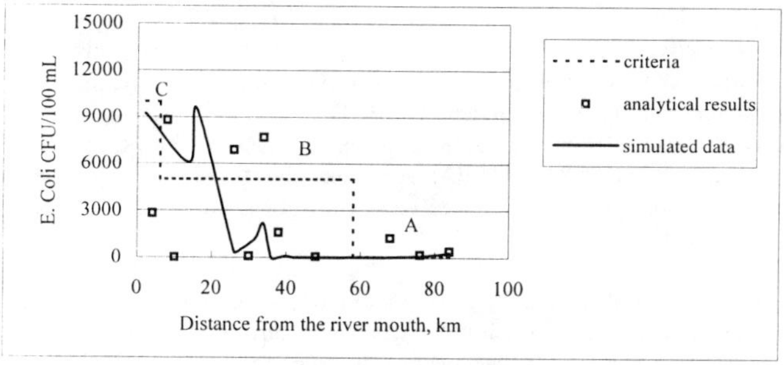

Figure 3. Measured and simulated E. Coli concentrations and three water quality classes (Classes A, B, and C) from the outfall (0 km) to the 84 km upstream location.

Effects of Foot-and-Mouth Disease Event on Kaoping River Water Quality

The hog population dropped to approximately 0.7 million in the whole Kaoping River Basin due to the occurrence of foot-and-mouth disease in 1997. Thus, the discrepancy of the water quality results between the years of 1996 and 1998 reveals the decreased impacts of hog wastes on water quality before and after the event of foot-and-mouth disease. The analytical results demonstrate that the 1998 water quality data are much better than the observed water quality in 1996. The analytical results show that the decreased manure from hog farming caused the significant drops in BOD and NH_3-N concentrations after 1997 (TEPA, 1998b; Ning *et al.*, 1998; Chiang *et al.*, 2000; TEPA 1999). This indicates that the high BOD and NH_3-N concentrations in the Kaoping River are mainly due to the discharges from hog farmings. However, the unchanged TP concentrations in 1998 indicate that hog wastes had limited contributions on TP loadings to Kaoping River. Results from previous studies indicate that the major causes of TP inputs into the river are domestic wastewater and NPS pollutants from agricultural areas (TEPA, 1998b; Ning *et al.*, 1998; Chiang *et al.*,2000; TEPA 1999 and 2000).

Carrying Capacity Calculation

The carrying capacity calculations for BOD, NH_3-N, and TP were performed using the calibrated QUAL2E water quality model to obtain the maximum acceptable BOD, NH_3-N, and TP loadings per day without violating the water quality criteria for Kaoping River (Table 1). The calculated BOD, NH_3-N, and TP carrying capacities were 27,700, 4,200, and 600 kg per day. The calculated E. Coli carrying capacity was 4.81×10^{15} CFU per day. However, the current BOD, NH_3-N, TP, and E. Coli loadings are almost 2.7, 9.4, 8.5, and 7.3 times higher than the calculated carrying capacities, respectively. To protect public health and improve the river water quality, seven river management scenarios are proposed. Tables 2a to 2d present the simulated BOD, NH_3-N, TP, and E. Coli loadings after the completion of each proposed plan. Because the raw water inlet of the water treatment plant is located at the upstream location of the Da-Shu Dam, Scenario 1 is to impose the hog ban in the upper catchment (upstream of the Da-Shu Dam). Due to the wide spreading of the hog farms in the upper catchment, only 6,000, 1,500, and 500 kg of the daily BOD, NH_3-N, and TP loadings could be reduced, respectively. Moreover, the application of Scenario 1 can only reduce 1.4×10^{15} CFU of E. Coli per day. It is noticed that one major pollution action is putting into practice by the Taiwan government agency to reduce the BOD and NH_3-N discharges into the river. A large subsidy program for removing almost half of the livestock husbandry in the upstream area was initialized in 1998 by TEPA. However, the simulated results show that the application of the subsidy program for livestock removal has only limited effects on the river quality improvement.

Results demonstrate that the following measures are required to reduce the daily BOD, NH_3-N, TP, and E. Coli loads to below the calculated carrying capacity: (1) hog ban in the whole Kaoping River basin, (2) sewer system construction to achieve 30% of connection in the basin within 10 years, (3) removal of 10 riverbank landfills, and (4) enforcement of the industrial wastewater discharge standards. Results indicate that approximately 48,000 kg of daily BOD loading can be reduced, and the remaining BOD loading (26,700 mg/day) is lower than the 27,700 mg per day BOD carrying capacity after the implementation of four river management measures. However, NH_3-N, TP, and E. Coli loads are still far beyond the calculated carrying capacities (Tables 2a and 2d). To effectively NH_3-reduce N, TP, and E. Coli loads, advanced wastewater treatment (e.g., advanced biological treatment, disinfection) technologies need to be applied in the wastewater treatment system.

Table 2a. Simulated BOD loading to Kaoping River after the implementation of each proposed plan.

Scenario	Measure	Reduced BOD loading (kg/day)	Remaining BOD loading (kg/day)
1	Hog ban in the upper catchment (upstream of Da-Shu Dam) to reduce half of the hog population	6,000	68,700
2	Hog ban in the whole basin	37,800	36,900
3	Sewer system construction (achieve 20% of connection in 5 years)	3,900	70,800
4	Sewer system construction (achieve 30% of connection in 10 years)	5,900	68,800
5	Removal of riverbank landfills	2,800	71,900
6	Enforcement of the industrial wastewater discharge standards	1,500	73,200
7	Scenarios 2 + 4 + 5 + 6	48,000	26,700 < 27,700 (carrying capacity)

Table 2b. Simulated NH_3-N loading to Kaoping River after the implementation of each proposed plan.

Scenario	Measure	Reduced NH_3-N loading (kg/day)	Remaining NH_3-N loading (kg/day)
1	Hog ban in the upper catchment (upstream of Da-Shu Dam) to reduce half of the hog population	1,800	37,600
2	Hog ban in the whole basin	17,300	22,100
3	Sewer system construction (achieve 20% of connection in 5 years)	50	39,350
4	Sewer system construction (achieve 30% of connection in 10 years)	100	39,300
5	Removal of riverbank landfills	800	38,600
6	Enforcement of the industrial wastewater discharge standards	-	-
7	Scenarios 2 + 4 + 5 + 6	18,200	21,200 > 4,200 (carrying capacity)

Table 2c. Simulated TP loading to Kaoping River after the implementation of each proposed plan.

Scenario	Measure	Reduced TP loading (kg/day)	Remaining TP loading (kg/day)
1	Hog ban in the upper catchment (upstream of Da-Shu Dam) to reduce half of the hog population	500	4,600
2	Hog ban in the whole basin	2,800	2,300
3	Sewer system construction (achieve 20% of connection in 5 years)	-	-
4	Sewer system construction (achieve 30% of connection in 10 years)	-	-
5	Removal of riverbank landfills	80	5,020
6	Enforcement of the industrial wastewater discharge standards	-	-
7	Scenarios 2 + 4 + 5 + 6	3,600	1,500 > 600 (carrying capacity)

Table 2d. Simulated E. Coli loading to Kaoping River after the implementation of each proposed plan.

Scenario	Measure	Reduced E. Coli loading (CFU/day)	Remaining E. Coli loading (CFU/day)
1	Hog ban in the upper catchment (upstream of Da-Shu Dam) to reduce half of the hog population	1.4E+15	3.4E+16
2	Hog ban in the whole basin	1.57E+16	1.97E+16
3	Sewer system construction (achieve 20% of connection in 5 years)	6.7E+15	2.87E+16
4	Sewer system construction (achieve 30% of connection in 10 years)	7.9E+15	2.75E+16
5	Removal of riverbank landfills	-	-
6	Enforcement of the industrial wastewater discharge standards	-	-
7	Scenarios 2 + 4 + 5 + 6	2.36E+16	1.18E+16 > 4.81E+15 (carrying capacity)

Conclusions

The water quality of the Kaoping River, particularly that of its downstream reach, has deteriorated markedly over the past 10 years. This has largely been due to (1) substantial increase of the hog population in the basin, (2) untreated municipal and industrial wastewater discharges due to urban and industrial expansion, (3) intensified agricultural development in the upper catchment, (4) NPS pollution caused by stormwater runoff from agricultural areas, and (5) riverbank garbage dumpings. Both point and non-point source pollutants are the major causes of the poor water quality. Thus, improvements of conventional wastewater collection and treatment, as well as reductions in the contaminant loads from point and non-point sources are required to improve the river water quality.

A comprehensive strategy for Kaoping River basin management has been proposed. The strategy consists of short-term management and improvement measures (e.g., 10 riverbank landfills removal), long-term structural measures (e.g., sewer system construction), and land use management and legislation (e.g., hog ban, enforcement of wastewater discharge standards, application of total maximum daily load program based on water quality simulation models). To reduce the NPS pollutant load discharged into the river during periods of heavy rainfall, best management practices (BMP) and targeted rainwater management systems need to be assessed and applied in the basin. After the implementation of the proposed measures, the water quality can be significantly improved and the BOD loading can be reduced to below the Kaoping River carrying capacity. To provide high quality drinking water for the residents in the Kaoping River Basin, two other issues have been addressed: (1) establishment of appropriate river and raw water quality criteria for the Kaoping River, and (2) application of the advanced water treatment technology for raw water treatment. Progress in these and other related areas would be essential if the challenges to river water management and water quality improvements are to be met.

Acknowledgments

This study was funded by National Science Council in Taiwan. Additional thanks to Prof. Chang N.B., Prof. Wen C.K., and Mr. Ning, S.K. of National Cheng-Kung University, Prof. Kuo J.T. of National Taiwan University, and Prof. Yu S.L. of University of Virginia for their support and assistance throughout this project.

References

Brown, L.C. and Barnwell, T.O. (1987). "The Enhanced Stream Water Quality Models QUAL2E and QUAL2E-UNCAS: Documentation and User Manual." Report EPA/600/3-87/007, U.S. EPA, Athens, GA, USA.

Chiang, P.C., Kao, C.M., Lin, T.F., Yan, Y.L. (2000). "Sustainable Taiwan 2011." Taiwan National Science Council, Taipei, Taiwan.

CPA, Construction and Planning Administration (2001). "Study on Quantitative Criteria of Affecting Factors, Delineation Guidelines, and Performance Indicators for Delineating the

Source Water Protection Area of Water Quality and Quantity for Water Supply.", Taipei, Taiwan.

CTC, China Technical Consultants, Inc. (1999). "Drinking Water Protection Plan in Kaoping River, Report to Kaohsiung County Environmental Protection Agency." Kaohsiung County, Taiwan.

CTC, China Technical Consultants, Inc. (2000). "Drinking Water Protection Plan in Ping-Tung County, Report to Ping-Tung County Environmental Protection Agency." Ping-Tung County, Taiwan.

Ning, S.K., Shu, S.Y., Chiang, C.L., Yang, L., and Chang, N.B. (1998). "Water quality analysis in Kaoping River basin, The 11th Environmental Planning and Management Conference." Tainan, Taiwan, pp. 216-223.

TEPA, Taiwan Environmental Protection Administration (1998a). "Water Pollution Act.", Taipei, Taiwan.

TEPA, Taiwan Environmental Protection Administration (1998b). "Investigation and Remediation Planning for Kaoping River and Tung-gan River.", EPA-87-G105-03-03, Taipei, Taiwan.

TEPA, Taiwan Environmental Protection Administration (1999). "Industrial Wastewater Management Strategy (I)." EPA-88-U1G1-03-117, Taipei, Taiwan.

TEPA, Taiwan Environmental Protection Administration (2000). "Industrial Wastewater Management Strategy (II)." EPA-89-U1G1-03-002, Taipei, Taiwan.

TEPA, Taiwan Environmental Protection Administration (2001). "Investigation of Non-point Source Pollution in the Drinking Water source Water Protection Area of Kaoping River Basin." EPA-90-G103-02-223, Taipei, Taiwan.

Modeling Watershed Protection Strategies for the Aurora Reservoir Watershed

T. Andrew Earles, Ph.D.[1] and Jonathan E. Jones, P.E.[1]

Abstract

Development in drinking water watersheds poses many water quality challenges. The level of water quality protection required goes beyond typical best management practices (BMPs) that are used for development in non-drinking water watersheds. To assess the treatment strategies that will protect the water quality of the reservoir two elements must be considered: (1) the allowable pollutant loading for the reservoir and (2) the loading from the development given a variety of treatment strategies. Water quality models serve as a tool for evaluation of both of these elements in the planning stages of a development.

This paper presents a water quality model developed for evaluating pollutant loading from development projects in the Aurora Reservoir watershed in the Denver metropolitan area in Colorado, USA. A major developer in the watershed, Laing/Village Homes (LVH), has worked closely with the City of Aurora (Aurora) to develop a water quality treatment system for the Southshore development that will limit pollutant loadings to the reservoir to a level that will protect the water quality of this drinking water supply. LVH has developed a model for evaluation of pollutant loadings from the development; Aurora is working to determine allowable pollutant loadings to the reservoir.

Keywords: Water Quality, Modeling, Reservoir, Watershed Management, BMPs, Ponds in Series, Protection Strategies, Stormwater Irrigation, Stormwater Bypass

[1] Wright Water Engineers, Inc., 2490 W. 26th Ave., Suite 100A, Denver, CO 80211, (303) 480-1700 (phone), (303) 480 1020 (fax), email: aearles@wrightwater.com.

Introduction and Background

The Aurora Reservoir and Watershed

The Aurora Reservoir is located in the metropolitan Denver area of Colorado, USA. The Aurora Reservoir was constructed in 1990 to serve as a terminal storage facility for the City of Aurora. In addition, the reservoir provides recreational uses including swimming, fishing and non-motorized boating. The normal storage volume of the reservoir is 3.95×10^7 m^3 (32,000 AF) with a surface area of 326 ha (805 ac) at normal pool depth. The maximum depth of the reservoir is 27.5 m (90 ft). The watershed surrounding the reservoir consists of roughly 1215 ha (3000 ac) of grazed, rangeland grass. Soils in the watershed are generally classified as sandy loams and are included in the Type B hydrologic soil group (Soil Conservation Service (SCS) 1971). Figure 1 illustrates the Aurora Reservoir and watershed as well as the LVH development parcel. The 324 ha (800-ac) development parcel drains to the reservoir through two primary arroyos, Senac Creek (West Tributary) and an unnamed eastern tributary (East Tributary). These arroyos are ephemeral and in many locations are deeply incised with steep, unvegetated banks up to 6 m (20 ft) high in some locations. Average annual precipitation is between 330 mm (13 in) (Parker gage) and 432 mm (17 in) (Stapleton gage).

Currently the reservoir is used primarily to supplement other Aurora drinking water sources during periods of drought and as a blending source for other lower-quality drinking water sources, resulting in a low current flow through rate of approximately 6.4×10^6 m^3/yr (5,200 AF/yr). This results in a residence time of over 5-yrs. Under build-out conditions, use of the reservoir as a drinking water source is projected to increase dramatically in response to increased demand, and the residence time in the reservoir is expected to decrease accordingly.

The Aurora Reservoir, unlike most drinking water supply reservoirs in Colorado, does not have state water quality standards in place (the Colorado Water Quality Control Commission has not yet established state standards). Rather, Aurora has developed water quality goals related to potential human health risks, taste and odor problems, water treatment considerations and Safe Drinking Water Act (SWDA) requirements. These goals, which are summarized in Table 1, were developed by Aurora's consultants in the late 1990's. To date, Aurora has not performed reservoir water quality modeling—the goals in Table 1 would likely be adjusted based on the results of such modeling.

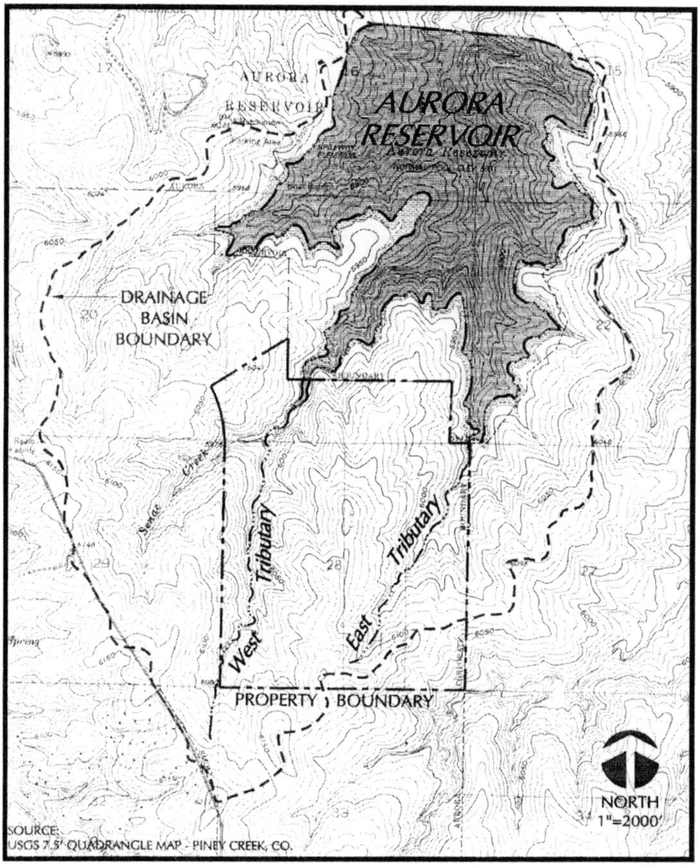

Figure 1: Aurora Reservoir Watershed (scale is approximate due to resizing of figure)

Table 1: Current Aurora Reservoir Water Quality Goals

Parameter	Proposed Goals/Thresholds for Action (Brown & Caldwell 1996)	Proposed Goals (CDM 1998)	Typical Reservoir Water Quality[1]
Nutrients and Organic Growth			
Total Phosphorus (μg/L)	30	8	10-30
Total Nitrogen (μg/L)	300	30	ND-70[2]
Chlorophyll-a (μg/L)[3]	15	5	0.23-2.33[4]
Secchi Depth (m)	>3.7 m (>12 ft)	>3.7 m (>12 ft)	2.1-6.7m (7-22 ft [5])
Dissolved Oxygen (mg/L)	none	>3	6.1-10.3[6]
Total Organic Carbon (mg/L)	4	----	3.2-5.9
Microbiological			
Total Coliform (#/100 mL)	----	----	0-1580 #/100 mL[6/7]
Other Parameters			
Fluoride (mg/L)	2		0.73-1.03
Sulfate (mg/L)	200		51-78
Chloride (mg/L)	100		27-30.8
Iron (mg/L)	0.1		0.03-0.19
Manganese (mg/L)	0.025		0.01-0.03
Arsenic (mg/L)	4		ND
Other SWDA Parameters	10 % of MCL		

[1] Average annual values based on up to 12 samples per yr, 1990-1998.

[2] Nitrate + nitrite, rather than total nitrogen.

[3] Growing season average.

[4] Values shown are range from four sites at two depths sampled during May to August during 1996 and 1997.

[5] Values shown are range from three sites sampled during eight-month period during 1997.

[6] Values shown are range from four sites at two depths sampled during eight-month periods during 1996 and 1997.

[7] > 10,000 #/100 mL sampled at swim beach; >100,000 #/100 mL sampled in West Tributary stormwater.

As the data in Table 1 indicate, the quality of the water in the reservoir is generally good. In 1996 and 1997, however, Aurora noticed declining water quality conditions with respect to nutrients, dissolved oxygen and algae (Note that many of the measurements reported in Table 1 for chlorophyll-a, Secchi depth and dissolved oxygen were taken during 1997). During 1997 and 1998 dissolved oxygen depletion extended as high as 15 m (50 ft) below the water surface of the reservoir, and the fall turnover of the reservoir resulted in taste and odor problems due to the mobilization of nutrients and metals, particularly iron and manganese, that occurred under the anoxic conditions.

A significant reason for this decline in water quality was a 97-mm (3.8-in) storm that occurred on July 21, 1997, producing a large sediment and nutrient load to the reservoir. Runoff sampled during 1997 from the undeveloped watershed had ortho-phosphorus concentrations exceeding 300 μg/L, nitrogen exceeding 500 μg/L and*E. coli* levels as high as 80,000/100 mL.

Because of the decline in water quality in 1997 and 1998, nutrients, especially phosphorus, have been of great concern to Aurora in considering proposed development in the watershed. The model described in this paper, therefore, has focused on phosphorus to address Aurora's concerns.

The Southshore Development

The proposed development, Southshore, is a 324-ha (800-ac) residential development with limited commercial areas. Approximately 3,000 homes are planned for the site. Approximately 295 ha (730 ac) of the development are tributary to the reservoir. The development of the parcel is planned in two stages. Initially, the West Tributary will undergo construction—this will afford an opportunity to evaluate the effectiveness of water quality protection measures, and, if necessary, strategies can be modified for the development of the East Tributary. Based on preliminary plans, the urbanization of the West Tributary watershed will result in an imperviousness of approximately 45% of the watershed.

The modeling presented in this paper has focused on the 150-ha (370-ac) West Tributary portion of the Southshore development. In addition, the model accounts for runoff from approximately 121 ha (300 ac) of offsite area draining through the West Tributary that is planned for residential development by others.

Given the development's proximity to the reservoir, the central riparian corridors through both the East and West Tributary watersheds and the presence of wetland pockets throughout the site, LVH has made it a goal to create a "water-oriented residential community." LVH has planned a network of open space, linking ponds, wetlands and parks—many of which will serve as both stormwater BMPs and aesthetic amenities. Stormwater reuse is also planned for the development—stormwater from ponds on the site will be applied for irrigation of parks and open space areas.

The Debate Over Existing Conditions

The pollutant loading under existing conditions is an important benchmark in the development of water quality protection strategies. The importance of this benchmark is magnified by the current absence of a lake water quality model. Aurora is developing such a model; however, project planning cannot simply be put on hold until the model is available. Unfortunately, data related to existing conditions is sparse, with only one runoff event sampled in 1997.

Estimates of existing conditions annual phosphorus loading from Southshore's portion of the West Tributary watershed have ranged from 8.2 to 40.8 kg/yr (18 to 90 lbs/yr) (Brown and Caldwell 1996, CDM 1998, Olsson Associates 2001). These estimates have been based on rainfall-runoff calculations and data for "typical" phosphorus event mean concentrations (EMCs) from the literature. For the Denver area, a "typical" EMC for phosphorus in runoff from an undeveloped watershed is 400 μg/L (Denver Urban Drainage and Flood Control District (UDFCD) 1999). The 400 μg/L phosphorus EMC is based on data collected as a part of the Nationwide Urban Runoff Program (NURP) in the Rooney Ranch watershed in the western portion of the Denver metropolitan area. Although the 400 μg/L EMC is based on the best available data in the area, it is far from ideal for the Aurora Reservoir watershed for a number of reasons:

1. The EMC is based on a very limited number of samples—two events. Further, these two events were relatively small. A larger amount of precipitation (the average annual event, for example) would likely have different pollutant mobilization characteristics.

2. The geology of the Rooney Ranch watershed and the Aurora Reservoir are dramatically different. Material present in the Rooney Ranch watershed tends to armor the channel; channel bottoms in the Aurora watershed have sandy bottoms and show no signs of armoring.

3. Channel characteristics differ significantly—channels in the Aurora Reservoir watershed, particularly the West Tributary, are deeply incised, with steep, eroding, unvegetated banks. Photograph 1 shows a West Tributary arroyo. In the Rooney Ranch watershed, channels are far more stable. There are few signs of degradation and in most reaches the channel and banks are vegetated.

4. Though both watersheds are undeveloped, there is little similarity in ground cover and vegetation. Photographs 2 and 3 show typical conditions for the Aurora Reservoir and Rooney Ranch watersheds, respectively. Ground cover in the Rooney Ranch watershed is generally good; ground cover in the Aurora Reservoir watershed is poor to moderate, largely a result of cattle grazing.

Given these factors, it is unlikely that a 400 μg/L EMC for phosphorus is appropriate for the Aurora Reservoir watershed. In fact, data from the single runoff event that was sampled in the West Tributary tend to corroborate this—the ortho- fraction of phosphorus alone exceeded 300 μg/L.

Since existing conditions load estimates range widely and since the basis for calculating the existing load is debatable, the "target" load for modeling is somewhat nebulous. For the sake of initial modeling, a load in the middle of the range of the estimates, 22.7 kg (50 lbs), was used as the "target" for pond performance.

Water Quality Protection Strategies

Basic Requirements

In the Denver metropolitan area, criteria for water quality protection measures have been developed by the UDFCD. The UDFCD criteria target smaller, more-frequent events for water quality. UDFCD requires provision of storage for the water quality capture volume (WQCV), which corresponds to the 80th percentile runoff event (Water Environment Federation (WEF)/American Society of Civil Engineers (ASCE) 1998, Urbonas *et al.* 1989). In addition to the WQCV requirement, UDFCD requires major drainage channels to be stable during the 25-yr event. UDFCD also has requirements for providing detention to reduce post-development flows to pre-development levels for frequent and infrequent events (up to the 100-yr event); however, since the Southshore development is immediately upstream of the reservoir, there is no flood risk to downstream properties from un-detained flows. As a result, this requirement applies to up-gradient developments (flood risk to Southshore) but not to the Southshore development—peak attenuation that is provided by Southshore will be for the purpose of water quality (to minimize channel erosion) rather than for flood control purposes.

Protection Plan

While the UDFCD criteria provide a high level of water quality protection for most development projects in the Denver metropolitan area, developments in drinking water watersheds require a higher level of protection. To provide a level of protection appropriate for a development up-gradient of a public water supply source, Southshore will employ a multi-tiered strategy consisting of structural and non-structural BMPs. Components of the protection plan include:

1. Providing storage in excess of the UDFCD WQCV, using a combination of retention and extended detention in a series of wet ponds and wetlands.

2. Potentially implementing a bypass system to draw stormwater from a terminal retention pond and convey it around the reservoir, discharging it over the spillway of the reservoir to a stabilized channel below.

Photograph 1: West Tributary Arroyo

Photograph 2: Aurora Reservoir Watershed (West Tributary)

Photograph 3: Rooney Ranch Watershed

3. Stabilization of major drainageways throughout the site for events up to the 25-yr event—the arroyos of the West Tributary are currently a major source of sediment and phosphorus to the reservoir.

4. Reusing treated stormwater for irrigation of open space areas and parks.

5. Implementation of water quality measures along major drainageways including drop structures, bio-engineered channels, etc... Arroyos on the site will be reconfigured to create greenways through the development, providing wildlife benefits in addition to water quality improvement.

6. Development of an integrated pest management plan (IPM) for large open space areas.

7. Water quality covenants for all residential and commercial property owners.

8. Public education emphasizing that the reservoir water is the same water residents drink.

9. Construction phase erosion and sediment control, emphasizing redundancy. The plans for construction call for construction of a terminal retention pond and stabilization of the drainageway as an initial task.

When properly designed and implemented, these elements of the water quality protection plan will provide a high degree of treatment for runoff from the development.

Spreadsheet Model

For the measures listed above to be effective, proper sizing is essential. Source controls will reduce the load to the system; however, the ponds (retention and detention) will be the workhorses of the system. Beyond the UDFCD criteria, there are no guidelines for sizing of water quality ponds. LVH recognized that the Southshore development requires a level of treatment beyond the UDFCD criteria, but the question of just how much storage needs to be provided was unknown. To answer this question, a spreadsheet model has been developed to evaluate the level of protection provided by a range of alternatives. While using a spreadsheet model is a bit more cumbersome and less computationally efficient than a numerical model compiled in FORTRAN or Visual Basic, this approach was chosen to avoid a "black box" approach and to make the modeling approach more readily understandable to Aurora.

Model Overview

Since the spreadsheet model was intended to allow for evaluation of numerous alternatives, flexibility was a key consideration in design of the model. The model performs calculations for hydrology, a specified pollutant (phosphorus) and residence time in ponds and consists of three modules: an offsite module (OS), the Southshore module (SS) and a down-gradient terminal pond module (TP). Each module has the capability to model up to nine ponds as detention or retention ponds and ponds and modules can be modeled in series by establishing links between worksheets and workbooks. In addition, the model has the ability to model consumption from stormwater irrigation and includes an option for a diversion. Figure 2 shows the conceptual framework for a module. The model uses a fifty-yr precipitation record from the Stapleton gage and operates on a daily time step. Due to computational limitations involved with modeling fifty yrs of hydrology and water quality on a daily time step, each module contains a separate spreadsheet for each yr, resulting in fifty linked computational workbooks for each module. Since spreadsheets must be opened and saved to update links and calculation time to update links is on the order of 5-10 seconds, running the model can require significant time and user effort (opening, saving and closing 150 files). To automate the process and make the model easier to run, macros were developed to perform the tedious process of updating links. Each module has a macro to execute the run, and the most upstream module (the OS module) contains a macro to run all of the modules sequentially.

Model Input

Hydrology

Since the core of the model is a hydrologic balance, data for precipitation; rainfall-runoff relationship; evaporation from ponds; evapotranspiration; baseflows; irrigation rates; diversion rates; and pond characteristics including permanent and active pool volumes, surface area and release rates are required. These inputs were parameterized as follows:

1. Precipitation—precipitation data from the Stapleton gage were obtained from the National Climatic Data Center (NCDC) (NCDC 2000). While the Parker gage is geographically closer to the reservoir, the period of record is not as extensive (20 yrs for Parker vs. 50 yrs for Stapleton). Furthermore, analysis by Colorado State University (CSU) indicated that there were significant gaps in the Parker gage data and some anomalies (dramatic changes in annual precipitation values) when the gage was relocated

in the 1980's. The Stapleton gage, fortunately, proved to be a more conservative source for input than the Parker gage since average annual precipitation for the Stapleton gage is approximately 100 mm (4 in) greater than the Parker average.

Given Colorado's climate, it was also necessary to consider snow events. Since runoff rates are considerably lower for snow events (sublimation occurs and the melting period generally extends over several days for large events), the equivalent depth of precipitation for snow events was spread out over a period of three days prior to runoff calculations. This likely still over-estimates runoff from snowfall events but, at least, is a conservative approach.

2. Rainfall-runoff—to generate runoff values based on the precipitation data, the Colorado Unit Hydrograph Program (CUHP) (UDFCD 2001). CUHP was applied for a range of recurrence intervals to determine runoff totals. Then a power function was fitted to determine an optimal equation for calculating runoff from rainfall.

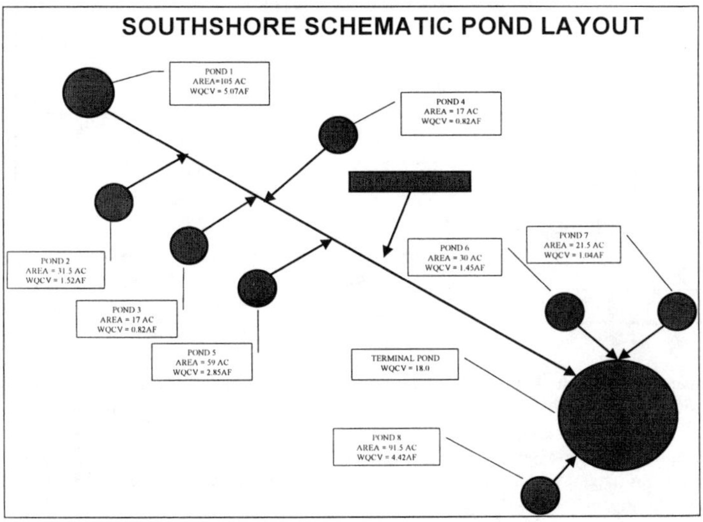

Figure 2: Conceptual Framework for Model Module

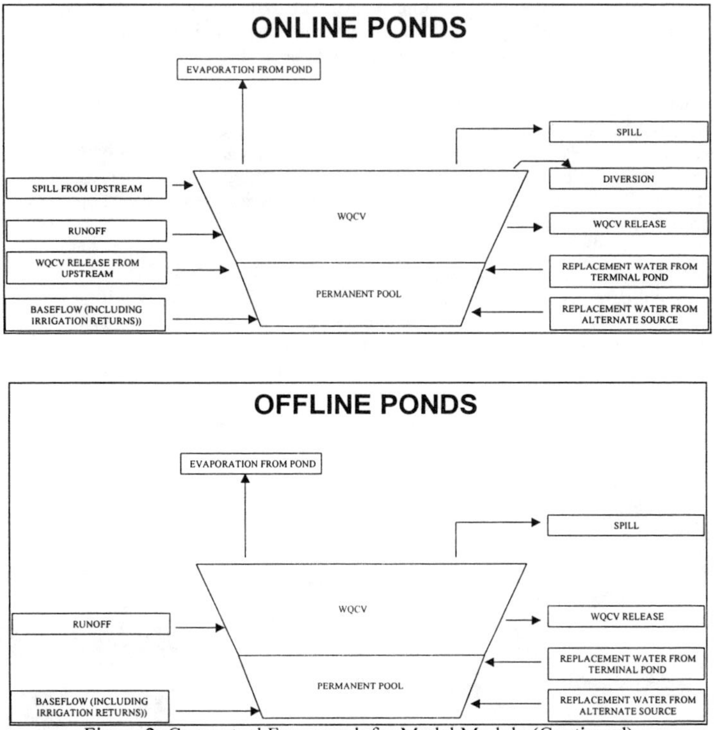

Figure 2: Conceptual Framework for Model Module (Continued)

3. Evaporation from ponds—pond evaporation was calculated based on average monthly values. The monthly values were evenly distributed over the days of the month.

4. Evapotranspiration—evapotranspiration (potential evapotranspiration (PET)) for areas requiring irrigation was calculated using the SCS TR-21 method (SCS 1967) using site-specific inputs.

5. Baseflows—baseflows, including irrigation return flows, were determined from analysis of data from other residential developments from the Denver metropolitan area. Based on data from seven residential developments, a strong relationship (R^2 = 0.95) exists between the imperviousness of a development and baseflows.

6. Irrigation rates—irrigation rates are a factor of the PET and the irrigation efficiency. The irrigation efficiency was a parameter that was adjusted in many simulations; however, a base efficiency of 80% was used for model runs presented in this paper.

7. Diversion rates—several diversion rates were examined: 0.11, 0.17 and 0.23 m^3/s (4, 6 and 8 cfs). These rates were selected based largely on the economic feasibility of implementing such a system.

8. Pond characteristics—the pond characteristics including permanent and active pool volumes, surface area and release time were the most manipulated parameters in the model. Permanent pool volumes were determined through consultation with the developer's site civil engineer (Martin & Martin); active pool volumes were varied in model runs in multiples of the UDFCD WQCV; surface areas of ponds were calculated based on a minimum permanent pool depth of 1.2 m (4 ft) (for aesthetic reasons); release times of 24 and 48 hours were modeled per UDFCD criteria.

Water Quality

Water quality input parameters include:

1. Post-development EMC concentration of phosphorus in runoff—a value of 400 μg/L of total phosphorus was used for this input. This value was based on UDFCD recommendations, data from the Shop Creek water quality ponds in the Denver area and data from the ASCE/USEPA Stormwater BMP Database (USEPA/ASCE 2001). The value for the phosphorus EMC was selected by mutual agreement between LVH and Aurora. The model addresses total phosphorus and does not explicitly model dissolved and particulate fractions separately.

2. Irreducible concentration—an irreducible concentration of 100 μg/L was used to set a limit on the achievable level of treatment. As with the EMC concentration, this parameter was based on data collected locally and data from the ASCE/USEPA Stormwater BMP Database. This is a conservative estimate of the irreducible concentration given the extended residence time of runoff in the system (in excess of 60 days for many scenarios) and primarily reflects removal by sedimentation of the particulate fraction of phosphorus rather than the biological treatment of the dissolved fraction that would occur with such extended residence times.

3. Removal efficiency—a system-wide removal efficiency of 70% was mutually agreeable to LVH and Aurora given the extended residence times, stormwater irrigation, source controls and use of BMPs in series. This removal efficiency is flow weighted between online and offline ponds. For example of all storage were provided by online ponds, the online pond efficiency would be 70%; if storage were evenly split between online and offline ponds, both types of ponds would have an efficiency of 45%, resulting in an overall removal of 70%.

4. Concentrations for baseflows and make-up water—a value of 100 µg/L was used for phosphorus in baseflows and make-up water (water needed during dry times to maintain a permanent pool).

Residence Time

Residence time calculations were an add-on to the model to evaluate the reasonableness of the removal efficiency assumption. Residence times are calculated in a similar way to pollutant loading—both assume that ponds (permanent and active storage volumes) are completely mixed and both use flow weighting to calculate a representative value for the entire volume stored. The residence time of the water in the permanent pool is an initial condition. The model was run for a variety of initial conditions and proved insensitive to this input—the model converged upon a stable average annual residence time regardless of the initial condition. An initial "age" of 30 days was used for water in the permanent pools.

Analysis and Discussion

Initially Aurora and LVH agreed to address a scenario in which LVH and offsite developments would provide 3 times the UDFCD WQCV in offline up-gradient storage (detention with a 24-hour release of the WQCV) and 3 times the WQCV in a large terminal pond (retention). Aurora requested the spreadsheet model to evaluate the reasonableness of this alternative and to evaluate other potential scenarios. Numerous scenarios have been modeled in the process of LVH's negotiations with Aurora staff. Table 2 summarizes several of the most insightful scenarios. Figure 3 illustrates the diminishing returns of increasing storage beyond the UDFCF WQCV. This was anticipated based on the work of Urbonas *et al.* (1989) who found that the volume of storage selected could be optimized at the point of diminishing returns by performing a regression between mean storm depth and storage volume to find the "maximized" WQCV. For the simulations in Table 1 and in all of the figures discussed below, a 48-hour draw-down was used for the WQCV. The model proved to be far more sensitive to the volume of storage and the presence or absence of a bypass than to the draw-down time—this is not a surprise given that the system modeled includes terminal retention ponds, which provide mean residence times in the system that greatly exceed the draw-down time.

Table 2: Summary of Modeled Scenarios

Model Name	Offsite Storage (Detention or Retention?)[1]	Southshore Storage (Detention or Retention?)	Bypass from Southshore Terminal (cfs)
3U3T	**3xWQCV** Up-gradient Detention **3xWQCV** Terminal Retention	**3xWQCV** Up-gradient Detention **3xWQCV** Terminal Retention	None
2U2T	**2xWQCV** Up-gradient Detention **2xWQCV** Terminal Retention	**2xWQCV** Up-gradient Detention **2xWQCV** Terminal Retention	None
1U1T	**1xWQCV** Up-gradient Detention **1xWQCV** Terminal Retention	**1xWQCV** Up-gradient Detention **1xWQCV** Terminal Retention	None
0.5U 0.5T	**0.5xWQCV** Up-gradient Detention **0.5xWQCV** Terminal Retention	**0.5xWQCV** Up-gradient Detention **0.5xWQCV** Terminal Retention	None
0.25U0.25T	**0.25xWQCV** Up-gradient Detention **0.25xWQCV** Terminal Retention	**0.25xWQCV** Up-gradient Detention **0.25xWQCV** Terminal Retention	None
1U2T	**1xWQCV** Up-gradient Detention **2xWQCV** Terminal Retention	**1xWQCV** Up-gradient Detention **2xWQCV** Terminal Retention	None
1U2T OS DETENTION	**1xWQCV** Up-gradient Detention **2xWQCV** Terminal Detention	**1xWQCV** Up-gradient Detention **2xWQCV** Terminal Retention	None
2U1T	**2xWQCV** Up-gradient Detention **1xWQCV** Terminal Retention	**2xWQCV** Up-gradient Detention **1xWQCV** Terminal Retention	None
1U2TB4	**1xWQCV** Up-gradient Detention **2xWQCV** Terminal Retention	**1xWQCV** Up-gradient Detention **2xWQCV** Terminal Retention	4 cfs
1U2TB6	**1xWQCV** Up-gradient Detention 2xWQCV Terminal Retention	**1xWQCV** Up-gradient Detention 2xWQCV Terminal Retention	6 cfs
1U2TB8	**1xWQCV** Up-gradient Detention 2xWQCV Terminal Retention	**1xWQCV** Up-gradient Detention 2xWQCV Terminal Retention	8 cfs

[1]For offsite detention, WQCV is released to Southshore Terminal over emptying time; for offsite retention, Southshore receives only flows when offsite spills occur.

Figure 4 shows an evaluation of several of the storage alternatives not involving a stormwater diversion, and demonstrates several important points revealed by the model:

1. The initially proposed scenario (3xWQCV in upper detention ponds and 3xWQCV in a terminal retention pond for Southshore and offsite areas (Model ID: 3U3T)) would easily meet the targeted annual average load of 22.7 kg (50 lbs). However, this much storage may not be necessary to reach the target—using 1xWQCV in up-gradient detention and 2xWQCV in a terminal retention pond (Model ID: 1U2T) also would meet the target with some room the spare. Likewise, the other alternative, using 2xWQCV in up-gradient detention and 1xWQCV (Model ID: 2U1T) in a terminal retention pond, meets the target.

2. A comparison of Models 1U2T and 2U1T provides some insight into the allocation of storage between up-gradient detention ponds and a terminal retention pond. On average, the annual mass reaching the reservoir is not very sensitive to the allocation of the storage; with respect to the volume of flow, however, there is greater sensitivity—$1.3x10^5$ m^3 (105 AF) for Model 2U1T vs. $1.2x10^5$ m^3 (98 AF) for Model 1U2T. This may not seem like a large amount in the context of the total spill volume; however, it is of the same magnitude as the evaporation make-up needs on an annual basis. Allocating storage to a lower terminal retention pond is therefore modestly more efficient.

3. Figure 4 also illustrates the differences between scenarios where offsite areas have a terminal retention pond (Model ID: 1U2T) and where offsite the terminal pond operates as detention pond, sending released flows to the Southshore terminal retention pond (1U2T OS DET). Retention of offsite flows is preferable to detention since it results in lower average annual spills and loads (22.2 kg/yr (49 lbs/yr) and $1.2x10^5$ m^3 (98 AF) for Model 1U2T vs. 24.5 kg (54 lbs/yr) and $1.55x10^5$ m^3 (126 AF) for Model 1U2T OS DET). While a significant difference exists with regard to average annual loadings, there is not a lot of difference during the wettest yrs, which are dominated by spills from ponds rather than release from detention.

Figure 5 presents an evaluation of various bypass alternatives for the scenario with 1xWQCV in up-gradient detention storage and 2xWQCV in a terminal retention pond (offsite treatment is equivalent to Southshore). Figure 5 demonstrates that a bypass can be extremely effective in reducing average annual loads to the reservoir, reducing the loading by a factor of 5 for the 0.11 m^3/s (4 cfs) bypass scenario. The bypass has a similar effect during wet yrs, effectively reducing the loading for the worst yr in the 50-yr period by a factor of nearly 2 compared to the same system without a bypass. Figure 5 also demonstrates diminishing returns in sizing if the bypass. By implementing a bypass system, the annual loading is reduced dramatically; however, increasing the size of the bypass only results in incremental improvements. For example, doubling the bypass capacity from 0.11 to 0.23 m^3/s (4 to 8 cfs) only results in a reduction in the average annual load of 2.3 kg (5 lbs), or roughly 1/3 of the total load. Note that in the scenarios where a bypass is modeled, the assumption is made that the bypass only draws from the Southshore terminal pond—offsite terminal ponds do not have bypass capability, so presumably some sort of cost sharing would be required to offset the high cost of a bypass.

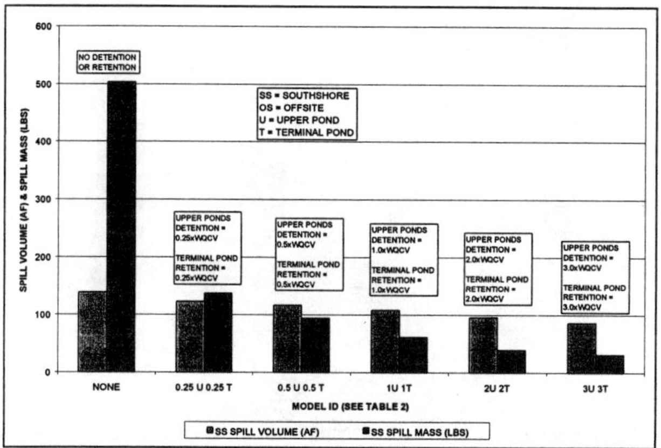

Figure 3: Illustration of Diminishing Returns for Southshore Spills (volume and mass) for Multiples of UDFCD WQCV

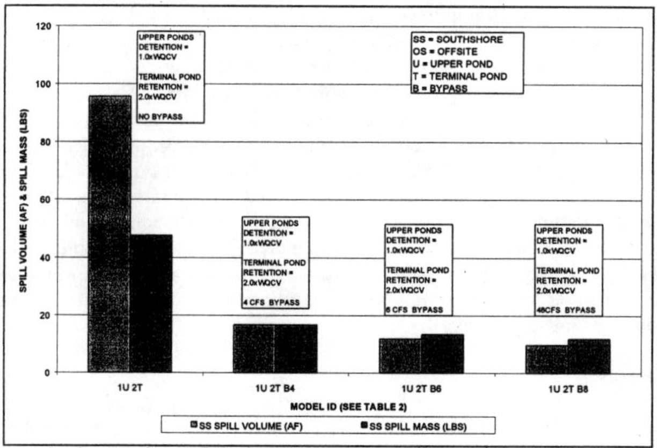

Figure 4: Comparison of Various Alternatives Without a Stormwater Bypass

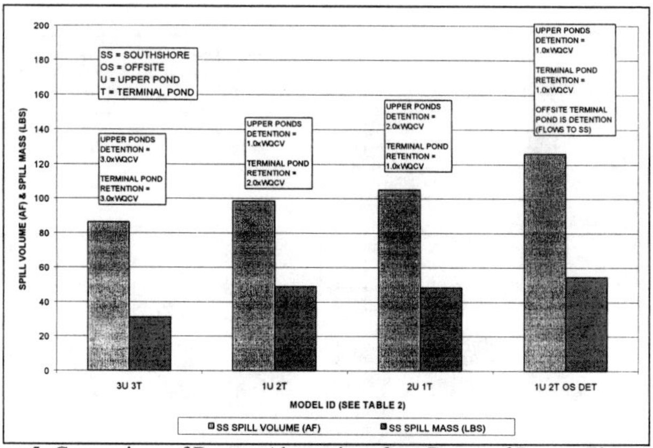

Figure 5: Comparison of Bypass Alternatives for a Range of Bypass Flow Rates

Conclusions

Development in drinking water watersheds poses many water quality challenges. The level of water quality protection required goes beyond typical best management practices (BMPs) that are used for development in non-drinking water watersheds. The spreadsheet model presented in this paper has proven to be a useful tool for evaluating detention and retention pond options for development in the Aurora Reservoir watershed in the Denver metropolitan area (U.S.A.). The model has allowed the developer and Aurora to examine pond-sizing alternatives, the impact using stormwater for irrigation, effects of offsite treatment strategies and bypass alternatives. Results of the model demonstrate that initial estimates of storage required to meet the targeted load on an average annual basis overstated the required size of the ponds—the average annual load target can be achieved with pond sizes site-wide reduced by 15,000 m^3 (12 AF) from the initial estimate. The model also demonstrated the effectiveness of a bypass at reducing loads, both for average yrs and for wet yrs. Using a bypass permits downsizing of ponds and provides other benefits as well including reliability in performance, easier stormwater management during construction, etc…

Acknowledgements

The authors wish to acknowledge several individuals who contributed to the work described in this paper. We would like to thank Matt Bachman and Jon Sorenson of Olsson Associates for their careful review of the spreadsheet model and their assistance in determining reasonable model input parameters. We would like to thank Dave Denton of LVH who has provided the financial support for this work. In addition, we have appreciated input from the City of Aurora during this project. Finally, we would like to thank Elizabeth Fassman

(University of Virginia) and Lisa Klapper (Wright Water Engineers) for their work on assembling the model and defining existing conditions.

References

American Society of Civil Engineers (ASCE)/United States Environmental Protection Agency (USEPA) 2001—*ACSE/USEPA Stormwater BMP Database.* USEPA: Washington, D.C.

Brown and Caldwell (B&C) 1996. *Aurora Reservoir Watershed Management Plan.* City of Aurora, CO.

Camp Dresser & McKee (CD&M) 1998. *City of Aurora Water Supply Master Plan.* City of Aurora, CO.

National Climatic Data Center (NCDC) 2000. *NCDC Summary of the Day, West 1, 2000.* Earth Info, Inc.

Olsson Associates 2001. *Correspondence with the authors via memoranda, letters, personal conversations and meetings from July 2001—to present.*

Soil Conservation Service (SCS) 1967. *Irrigation Water Requirements: Technical Release No 21.* United States Department of Agriculture, SCS, Engineering Division (Revised September 1970).

SCS 1971. *Soil Survey.* United States Department of Agriculture, SCS.

Urban Drainage and Flood Control District (UDFCD) 1999. *Urban Storm Drainage Criteria Manual, Volume 3.* UDFCD: Denver.

Urban Drainage and Flood Control District (UDFCD) 2001. *Urban Storm Drainage Criteria Manual, Volume 1.* UDFCD: Denver.

Urbonas, B.R., Guo, C.Y. and L.S. Tucker 1989. *Optimization of Stormwater Quality Capture Volume.* In *Urban Stormwater Quality Enhancement—Source Control, Retrofitting and Combined Sewer Technology.* American Society of Civil Engineers: NY.

Water Environment Federation (WEF)/American Society of Civil Engineers (ASCE) 1998. Urban Runoff Quality Management, *WEF Manual of Practice No. 23/ASCE Manual and Report on Engineering Practice No. 87.* WEF: Alexandria, VA.

The Strategy of Watershed Management and Public Participation

Li-Teh Lu[1]

Abstract

Public participation has been one of the key factors for the success of the watershed management. The conflict between watershed management and basic right of local people has proposed the important issue of establishing framework for public participation in watershed management.

China Time's ten-year experience of promoting planting in the watershed area was analyzed for time and module for public participation. Here emphasis is on practical experience and strategy.

Keywords : watershed, water resource quality and quantity protection area, planting 2000 trees to save Taiwan's water resource, public participation

Introduction

Taiwan area is specified with its small area and high population density. Due to specific weather and geographic factors, there is abundance of rivers, which could be divided into 129 river systems. There are 21 major rivers and 29 sub-major rivers. All these occupy 28429 square kilometers, which are about 80% of total Taiwan area. According to the Portable Water Zoning Plan, water resource quality and quantity protection area occupies 990,000 hectare, which is about one-forth of Taiwan area. According to the Water Use Act , the reservoir watershed area is about 360, 000 hectare.

Regarding Taiwan's topography, the length between north and south is longer then west and east. The central mountain areas have corsesed north to south and most rivers water runs western and eastern direction. With slope among 1/15 and 1/295, most of the major rivers are short with fast running water. The average year rainfall in Taiwan is 2515 mm, in which 74% becomes runoff. Most of rainfall comes from storm, which could occupy up to 70% of year rainfall. Due to steep slope of the most rivers and storm rain becomes runoff, rivers have poor water containing ability, which leads to the significant difference between dry and

1.Executive Secretary, Task Force of River Protection, the Chinatimes Cultural& Educational Foundation
Ph.D. Student, Graduate Institute of Environmental Engineering, National Taiwan University, Taipei 106, Taiwan.

wet seasons. The average rainfall for wet seasons (May to October) and dry seasons (November to April) is 7.78 and 2.22, respectively. This difference in significant in the Southern area, the ratio could be 9:1 and in the northern area could be 6.2 to 3.8. Besides, upstream watershed is located at new rocky shell, which is voluntary to water erosion. Due to deep slope of mountain area, storm will always make soil erosion dramatically, which leads to heavy sand sedimentation on rivers.

Due to concentrated storm in few months, and deep slope rivers, it makes the watershed management especially important. In addition, according to Portable Water Act, the water quality and quantity protection area occupies one-forth of total Taiwan area. The stress of the watershed development is under pressure. The beetle nut planting areas in 1996 was about 56,581 hectare. From 1971 to 1996, within this 25 years, the beetle nut planting areas increased more then 34 times. People plants beetle nut up to the mountain area, and 10,000-hectare is located in overdeveloped mountain area. In Taiwan area, more then 900,0000 hectare water resource protection area is designated as national forest area, one-forth of the area is private owned, and one-fourth of the area is designated as local government owned property. Within national forest areas, 120,000 hectare is belonging to aboriginal reserved area, and part of the area is rent for planting. In another words, more then half of the water quality and quantity protection area is belonged to aboriginal, rent forest planting, and private sector that leads to the conflict between the management operation and private benefit. Before the resolution of ensuring right of private sectors and executing government authorities, it's a strict challenge for watershed management. To involve the public participating in the watershed management and resolved the conflict has becomes an important issue.

In general, local people prefer developing then protecting in watershed protection area, but non-local people prefer protecting. This paper review China Time's experience and effort in promoting planting trees in watershed area. The analysis of the time and module of the public in watershed was conducted to provide practical experience and strategies.

The Case Study of the Public Participating in Watershed Management: Introduction of the Activities of Planting 20 Million Trees in Taiwan Watershed

In recent year, the nature disasters happen frequently. It often happens to be drought or flood and impact the lives and properties of the public. Investigation the cause, besides the nature disasters, improper development and utilization of mountain area is highly correlated. Overdeveloping in dam watershed, over-deforesting and bed water-soil conservation will lead to heavy soil erosion. How to manage watershed properly and sustainable utilization of water resources have been becoming an important environment issue.

From 1992, the China Times Foundation got together the China Time, Environmental Protection Administration, the Ministry of Education, The Commission of Agriculture, The Bureau of Forest, Nature Ecology Conservation Association of the Republic of China, the Mountain Respecting Foundation, and the Mandarin Times, , etc., about 35 public and private communities, to promote "Planting 20 Million Trees to Rescue Taiwan Watershed." It has been ten year of promoting the civil participation in planting trees to rescue the water resource.

1992 Activities on "Planting 20 Million Trees to Rescue Taiwan Watershed"

In 1992, the China Times got together the Environmental Protection Administration, the Ministry of Education, The Commission of Agriculture, The Bureau of Forest, Nature Ecology Conservation Association of the Republic of China, the Mandarin Times, Omen Public Relation Ltd., the China Petroleum Company, Sho-Shan Cultural Foundation, the Da-Jar Stream Ecological Environmental Maintain Association, Der-Jeor Law Firm, and Ching- Jei Accounting Firm to plan the project of one-person one-tree. On March 8, April 19, and June 21, 1992, in Taipei, Taichung, and Kaohsiung, respectively, activities of adapting one tree for one-hundred NT dollars of "Planting 20 Million Trees to Rescue Taiwan Watershed" were conducted to strengthen promulgation. The number of participating public and private groups is up to 35. At strategic side, it adopted the principle of "Think Taiwan All Around, and Act locally." Integrating the local characterization, these activities were designed and conducted. To deliberate the proper utilization of nature resource without destroying ecology balance, the Don-Shan River Recovery Program were designed as an example, and the environmental education tour of the Don-Shan River was conducted. In middle area of Taiwan, Da-Jar River ecological environment was suffered destruction. The Da-Jar Stream Ecological Environmental Maintain Association made effort to drive local people's attention and to involve more people to practical participating in action and conservation. In No. 22 Forest, Huang-Liu Stream, and the planting trees activity and environmental interoperation and larval rehabilitation were conducted. In the southern area, the cooperation with the Buddhist Association of paper recycling, the recycled disposal paper up to 37,900 Kg, which were sell at the price of NT 74,247 dollars and this fund were used as planting tree foundation. All these activities had contributed to the purpose of planting trees directly and indirectly.

1993 Activities on "All Birds Going Back to the Forest and All Stream Going Back to Ocean"

The activity them of 1993 was named as "All Birds Going Back to the Forest and All Stream Going Back to Ocean." The activities included sponsor recruitment, family nursing plant and platform, giving junior plants, and sponsors representatives planting in watershed.

To accompany the collected fund form the public, in the watershed of the Fei-Tsui Dam, the Wu-Shan-Tou Dam and the Fung-Shan Dam were planted up to 91.5 hectare in 1993. In principal, the budget for planted land and planting were planned by the government agencies. The public could contribute money for nursing junior plants.

To let sponsor realize the utilization of the fund, the China Time made a feature article to report the activities in 1992, the achievement of planting trees and its vision. That was also mailed to the part of sponsors for their reference.

The main stream of 1993 activities is to recruitment the sponsor. March 6, 1993, the organizer arranges 50 sponsor representatives to visit the result of planting trees in 1992. The participating business and communities representative included Cheng-Long Ltd., Taipei Normal School, Chen-Tai Ceramic Company, Weelpool Technology, ….,etc.

On March 12, 1993, the holiday of Planting Trees, a special speech "Green the Land, and Purify the Human Mind," were given by Master Sheng-Yen This was arranged for awaking the concept of co-prosperity among people and people, and environment. It was hoped that planting trees to recover the original green to let Buddhist "the clean land" come earlier.

In Kaohsiung, water resource degradation had always been a nightmare. Considering that, planting tress to rescue watershed activities were focused in 1993. In Fung-Shan Dam, 156,000 trees were planted to conserve the most important watershed in big Kaohsiung area. The people in southern area were actively participating in the private sectors' largest-scale planting activities.

1994 Activities on "Planting a Appreciating Tree for Mother"

Land is mother of all creatures. Without land, creatures could not grow. The third year of "Planting 20 Million Trees to Rescue Taiwan Watershed" activities, the theme in 1994 was "Planting a Appreciating Tree for Mother." This theme indicated the intention to awake more people's appreciation to all creatures' mother through mother's love into the task of planting tree to protect the earth. February 19, 1994, a press conference was held to inaugurate that year's activities. On March 6, an announcement activity, "Planting a Appreciating Tree for Mother", was held in Da-Chu-Ken Mountain area, Ray-Fan, Taipei County. In March 7, 1994, promulgation activities were held at Chin-Hu Secondary School in Chin-Jue County and Chi-Wen Primary School in Mew-Lee County, respectively.

From March 6 to May 8, the Mother's Day, in 1994, there were up to 17 participating counties, including: Taipei County, Hsin-Chou County, Mew-Lei County, Kaohsiung City and County, Tai-Chung City, Nan-Tou County, Jan-Hwa County, Yun-Lin County, Cha-Yei City and County, Tai-Nang County, Yee-Lan County, Hwa-Lein County, Tai-Dong County, and Peng-Hou County. They employed the parks, park designated or school designated area as planting area. These planting activities were employed to express the appreciation to their mothers, and these activities were jointed with environmental education activities, such as paper recycling.

In first two-year, the focus of "Planting Trees to Rescue Watershed" was on planting trees on watershed. To avoid watershed destruction, the public was not encouraged to visiting watershed during the whole activities. This also prevents the practical participation of the public. The activity of "Planting a Appreciating Tree for Mother" in 1994 was designed to break this restriction and integrate watershed planting trees and urban area greening activities.

The organizer pointed that the activity was to integrate urban greening and watershed tree planting. Based on the sponsors' donation, part of sponsor could have opportunity to visit the location of "Planting an Appreciating Tree for Mother" and to plant a tree for their mothers. All donation form the public was used in watershed planting tree related activities.

It was acknowledged that the China Forecast Company held "Remember" concert in Sun Yat-Sin Memory Hall three days before and after the Mother's Day. The bonus of this activities is about NT 1, 118, 800 dollars was contributed to the Planting Foundation. Besides, sponsor recruitment activities were held on May 21 and June 4, 1994.

1995 Activities on "Greening the Land and Treasuring the Watershed"

Considering the topography, geology and population density of Taiwan area, dam development could solve the problem of water usage. However, the forest protection in the watershed could ensure water quantity and quality, and extent the life of the dam. In 1995, the "Rescue Fei-Tsui Dam Activity" was pronouncing at the Planting Holiday. In Chung-Jen primary school, Hsin-Deng County held an nursing junior plants, yang students participating in the opening ceremony, and visited Fei-Chui Dam to plant tress there. This practical participation and protecting the Taipei's most important water supplier, Fei-Tsui Dam.

The focus in 1995's activities was on rescuing Fei-Tsui Dam. Due to government's taking the rent forest area back, we needed to plant the area immediately to prevent overcutting trees. According to Taipei's budget in Taipei Municipal budget planing, the total budget is about NT 4, 000, 000. Base on the scale of budget, we might need 70 years to planning 1,000 hectare's forecasting programs. It seemed too long.

The Fei-Tsui Dam is one of the most important water rescue suppliers, it bothersome that we need 70 year to planting trees. How to involve the public in this activity became an important subject. Therefore, in the morning of March 12, 1995, a simple ceremony was held at Hsin-Den Primary School, kids made a puzzle of "Rescuing water resources of Fei-Tsui Dam Fei-Tsui Dam" to response this activity. They made signature on the puzzle and their practical involvement to planting and nursing junior plants in Fei-Tsui Dam, and then move the junior plants to the watershed areas to rescue the Dam.

1996 Activities on "Farewell to Drought and Rescue the Water Resources"

From 1994 to 1995, Taiwan faced with drought. Even in rainy harbor, Keelung, the drought phenomena happened. The need for conserving forest to preventing drought became more important. Based on the previous two-year drought, the theme of activity in 1996 was developed as designated as "Farewell to Drought and Rescue the Water Resources."

The first activity in 1996, the public hearing of the Fate of Okinawa Pine Forest in Fei-Chui Dam, was held at the Legislative Yuan. Due to nematode disease of Pine tress along the Pei-I highway, a lot of pine trees died. Although it is important to plant tree to rescue watershed, the nematode were spreading and protection of these pine became significant important. Regarding this theme, agriculture, forest, plant pathology experts were invited for brain storming and developing the resolution.

In March 17, 1996, around Two-Tang section of Fei-Tsui Dam, the planting activity of "Planting trees to rescue watershed, and sending love to Fei-Tsui Dam" was held. More then 400 people participated in this adapting tree activity.

There is one kindness people, who lived in Shi-Lin, Taipei, donated silver coins, which is valued more then 4 million dollars. These silver coins were sailed by the China Time, all the revenues was donated to Fei-Chui Dam for planting foundation. This had been the largest amount of money received since 1992. Besides, the Rotary Club, Young-Ho, donated NT 650,000 dollars to the activities of planting tree to rescue watershed. After few years of activity promotion, the planting trees to rescue watershed tended to be more diversified. Beside the promulgation in different areas, the TV forecast also involved in the promulgation activities. On June 29, 1996, "Planting trees to rescue watershed, and sending love to Dam. Thanksgiving Eve" was held in the Chung-Lei Culture Hall. The program was on TTV on July 7, 1996. Another "Happy Fei-Chui - planting trees to rescue watershed, Thanksgiving Eve" was held on June 30 in the China Time Tower, the program was on TTV on July 6, 1996.

1997 Activities on "Seeding hope to farewell wind and flood disaster"

From June 31 to October 1, 1996, typhoon Herb impacted all Taiwan area. Taiwan faced the most serious wind dizaster and flood. Especially in the middle mountain area there were heavy land slide with mud and rock. Overdelopment of the forest area induced the strong response of the nature. If we could promot the planting tree activities, then we could save our land as well as all Taiwan people's lives and properties. The lesson of typhoon Herb showed the importance and the vision of the activities of Planting 20 Million Trees in Taiwan Watershed initiated in 5 years before.

In the previous 6 year, the China Time noticed a report in Norway. A primary school initiated a rescue activity to adapt rain forests in Brazil. Under the effort of these children, the kids all over Norway participate this activity. They saved their money for buying toys or candies and donated to buy a rained forest which had bee sailed for lumber usage. This story drive the China Time to get together schools, the Mandarin Times, Environmental Protection Administration, the Ministry of Education, and the Taipei County Government to promote the activities of "Seeding hope to grow up with junior trees."

From 1995, "grow up with junior trees" had brewed the affiliation of children in school to take care of trees. The organizer consulted with the Ministry of Education for many times on the involvement the activities into their contest programs and consensus was reached.

In March 1997, Post Office Bureau published "Protecting Watershed Stamps" in NT 5 and 19 dollar-face values. For 19-dollar stamp, there was tree and river all together that showed the close relationship between planting tree and watershed.

In 1996, new poet contest was held and the theme was designated as "the impression on the activities of Planting trees to rescue watershed" and the "how you feel about the importance and description of the watershed and forest." That activity lasted for one month and a half, and 800 articles were participated in that contest.

Besides, in Cha-Yei county, "Farewell to wind and flood disaster" planting tree activity was held. All county citizens in Cha-Yei and Taipei with tree in their name were invited to participate in. Another "Farewell to wind and flood disaster" planting tree activity held in Dong-Shi, 500 primary school student and the public participated in.

Promote Tree Melody and 9-Year Planting Tutoring Material

After few year of promotion, the module of government and private cooperation in planting tree had been establish. The continuity would be an important issue in the future.

In the previous activities, we found that there is one primary school course regarding observation of bean sprouting. We employed this course to develop "Growth Up With Bud" activities. We hope to integrate the vision and action of planting trees into the normal education programs. The nature science of the 2nd and the 3rd grades were added lectures of brewing buds. The irrigation, observation and recording were conducted by students. When bud grew up to tree, it will be transplanted in the garden of the school, community or watershed. To implant their love to tress form yang ages.

To sustain the whole activities, especially in grade 2 and 3 of the primary schools' bean sprouting course, the facilitation of children to plant seeds were conducted through daily lecturing activities. Each student were taking care of 10 seeds, sprouting, and planting to watershed, school, or community to achieve the target of one people one tree. The whole program was mainly conducted through normal education, and integrated family education to implant the vision of planting by all citizens.

Being facilitated by Environmental Protection Administration, Professor Chou, Zue of National Normal University, and Professor Chen, Pei-Pei, we employed the ideal of "Grow Up with Bud" to develop a set of planting facilitating tutoring material. We integrated the care for forest, trees, and environment and introduce and promote these ideas consequently. It was hope that the education and knowledge would act as root for action. Let children understand the importance of forest for soil conservation, and devoted themselves to the task of planting trees and protecting forests.

Turn Betel Palm into Green Woods

In 1999, 921 earthquake happened to damage the topography of Taiwan area heavily. The area of landslide at Jue-Fan 2nd Shan in Nan-Toe County covered hectares. After earthquake, citizens in

In the Chung-Lia town diverse management, the upstream watershed of the Mou-Low Stream, people felt the importance of rebuilding eroded mountains. They realized that the disaster induced landslide as well as impacted their economic resource, beetle nuts. They were conscious that forest recovery should be the resolution. A voluntary initiation "Beetle Nuts to Green Forest" was developed to recover the eroded mountains to be green forest. Containing water resource and soil was the vision for sustainable management of Taiwan. The China Time Foundation initiated the sponsorship program of "Beetle Nuts to Green Forest." Local private sectors, such as the Taiwan Plastic Cooperation, Chung Rong Fa Charity Foundation, Taiwan Tea Cooperation, Ren Bou Computer Cooperation, Mandarin Commercial Bank Cooperation, Rey Hwa Investment Cooperation, Da Hwa Stocks Ltd., Yee-Mei Environmental Protection Foundation, and China Steel Cooperation, were invited to participate in the program. Up to now, more then 24,140,000 dollars were collected and the cost for planting activity was up to 36,110,000. "Beetle Nuts to Green Forest – Welcome the First Millennium Light, Plant A Tree for Upcoming Century" was introduced to plant trees for next generation.

The Achievement of Activities of Planting 20 Million Trees in Taiwan Watershed

1. Adapting planting trees in watershed, the main stream of activities of planting trees to rescue watershed. The public could use one hundred to adapt a tree; the money could be donated through the post office system or CPC system. All the donated money will be

employed in the task of the activities of planting trees to rescue watershed. Up year 2000, all funds received, 61, 320,000 dollars, were invested in watershed planting activities.

2. There were more then 200, 000 buds were planted in watershed area.
3. The whole activities were focused on Fung-Shan Dam watershed, and more then 156,000 trees were planted. It occupied one-forth of the whole area of Fung-Shan Dam watershed.
4. Promoting planting forest activities in Fei-Tsui Dam watershed. It covered about 1000 hectare and 2,000,000 trees were planted.
5. In Chung-Liao town, 123,000 trees were planted to substitute Bette nut trees.
6. In accordance to 9-year obligation education program, tutoring material had been developed and " take care water and protect forest education program" were promoted.
7. To urge government to promote activities of "Whole Citizen Champaign of Planting Forest" and "One People One Tree – Hold Land with Root of Trees", etc.

Models of public participating in watershed management

People participate in watershed management on upstream using many methods. The method we discussion is to plant. In general, uncovered land cannot reserve water effectively. Some experiments show, in forest there is 25% rainfall placed on tree-crown and then is vaporized. 25% rainfall runoff slowly and 25% supply to ground water. Another 25% stays in soil. Nevertheless, rainwater dripped down on uncovered land. The 50% rainwater runoff fast and 40% is vaporized from ground surface. To have only 10% rainfall supply to groundwater. Therefore the capability to retain water for forest is larger 5 times than uncovered land.

Forest is also contributive for soil conservation and preventing flood. The test of Tennessee River in U.S.A. showed that the flood discharge decreased to 1/38 when they plant one hundred thousand trees in Tennessee River basin. Afforestation in large watershed can eliminate the soil washout. In small watershed, the silt discharging to river decreased 90% after afforesting 3 to 4 years.

Reference to the research of Yi-An Ren in forestry testing department in Taiwan, the forest can contain rainfall, prevents floods, increases the soil porosity and resupplies groundwater. The forest watershed area in Taiwan is about one million five hundred seventy thousands hectare. If the forest is in good condition in Taiwan, the forest can store 1.4×10^{10} m^3 water resources that is larger 7 times than domestic water of one year. Therefore planting is important refer to conservate water resources.

Public territory：Model of Jadeite Reservoir

Assignments of afforestation in watershed of reservoirs must be executed by government. Because the budget of government is limited so the afforesting budget may be delinquent. For example, Fei-Tsui Dam was operated in 1970s and government expended nine hundred forty million dollars NT to compensate for the private goods in the watershed. Total retrieve public forest area is one thousand hectare below altitude 171 m. The total cost for afforesting is about two hundred eighty million but the Taipei municipal government just prepare budget four million dollars NT per year. Therefore the action of 「Planting twenty million trees to save Taiwan water resources」 exposed the revelation that municipal government neglected the afforestation and then originated the activity 「One people one tree save Jadeite Reservoir」. The first step public participated in fifty hectare afforesting. Then public voice came into existence. The second step: coordinate with parliamentarians to held a public hearing「Seventy years needed to recover Fei-Tsui Dam forest ?」. At last inviting Taipei city Mayor Shui-Bian Cheng attended this meeting and then he promised finishing afforestation at least five years ago. This is a successful case that public and government participated in watershed planting.

The cases of public participation are such as Taipei Chin-May Lion Association planting one thousand trees in Jadeite Reservoir watershed and Taipei Da-Dou-Cheng Lion Association, Taichung W-N Lion Association; Tainan Center Lion Association also participated in afforestation. Another joined groups for instance International Rotary Club 3490 area second group and fourth group (included Taipei Bang-Chou North area, Center area, East area, South area and West area Teams, New Zealand International Rotary Club, Japan International Rotary Club and Taipei Yan-Ping Rotary Club). The others are Cathy Airline of Taiwan and Singapore Airline participating in afforestation.

The methods of public participation are 100 dollars NT per one Tree and supporting one thousand trees by groups. The sponsors can name the planting area, set memorial steles, plant on the spot and visit the sapling growth.

For expending the public participation in the activity of「One people one tree save Jadeite Reservoir」, the unit of planting is half hectare that we invited the employees of any companies and corporations to plant. The sponsors denominated their planting areas by names of their companies. For example, the staffs of Taiwan Agriculture Department are about one hundred people presence to plant and leading man is the vice-chairman of Agriculture Department. They plant two hectare at least two hours ago and dominated as "Agriculture Department Staff Forest". This was the first forest named by corporation name in 1995. The Forever corporation of U.S.A. contributed three million and dispatched staff to afforest then dominated "Forever Co. Memorial Forest".

Private Owned Land : Model of Her-Hsin Village Sufferer

There were about 200 families lived in the Her-Hsin Chuern waster source area. More than 1500 acre land was covered mainly by betel palm and spring banana trees. Because of the damage by 921 Earthquake, about 110 acre area was slided and about 500 acre land was over utilized. Because of the potential risk to the safety of the residents of Her-Hsin Village, Fu-Sen Village, Ba-Sien Village, Jon-Liao, Da-Tze-Wan, and Nan-Tou City and to the sanitation of drinking water and irrigation, the villagers initiated the "Salvation and Afforestation Campaign". Chinatimes Foundation echoed to their objectives and worked with the Rebuilding Committee of Executive Yuan to rebuild Jio-Fern-Er Mountain and Nan-Tou Jon-Liao: "Turn Betel Palm into Green Woods" Adoption Project. The meaning of the activity included:

1. Shocked by the damage of 921 Earthquake, the residents self-realized the seriousness of over planting betel palm and destroying the landform. They determined to afforest to save water source and to save their land.
2. The suffered people were facing the hardship of jobless. This project could use the sufferers and help them away from difficulty.
3. The over used and slided land in Taiwan was more than 10,000 acres. The success of

landform recovery of this project could be used as a demonstration, and extended to other area.

4. Because of the climate change and the fast deforestation, flood would happen frequently. It is necessary to afforest to protect the land and water source.
5. Forest is the common resource of the public, it also can purify the atmospheric air and improve the climate. The project land was designed as an "Organic Culture Village" to incorporate forest tourism and organic agriculture to develop a new agriculture model with a balance of environmental protection and human existence.
6. The first phase was planned to plant 120,130 trees on 60.07 acre land, which was provided by 74 farmers. The second phase was for over-used 500 acre land, which was able to be adopted for afforestation.

This Campaign began from January 1, 2001 at Jon-Liao Hsian, NanTou County to "Receive the First Aurora of 21 Century and plant the First Tree of the Century". Representatives from several Sponsors, such as Taiwan Integrate Circuit Foundation, HwaBan Electronics, Compal Electronics, Inc., Evergreen Group, and YiMei Environmental Protection Foundation, were invited to participate the opening. Besides, Her-Hsin Chuern carried on a "Tour of Healthy Life" program to help the sufferers to walk away from the disaster. Meanwhile, in the first thirteen weeks, each adopting enterprise would encourage her employees to help and adopt the project for one week, by joining the eco-tour to afforest and to help the economy in the damaged area.

Model of Aborigine participation

Aborigine was the population who concerns the issue of environment protection the most. Land of the aboriginal reserve area could be provided by landowner to be reforested after encouraging by appropriate feedback while the land opened to be possessed.

They were planning to promote grouping old people around every tribe to join the activity o f afforestation. It was designed to use the ethic of esteemed older and pundits to promote the activity of afforestation overall.

There was 12 acres of aborigine area around Taiwan. Over 12,993.4 acre land was over used. Only in Jen-Ai county, the over used land was about 1/3 of total over used. The aborigine area was also facing great pressure of developing. To reduce the pressure from developing, lawmaker WaLis Balin and Committee of Aboriginal Concert Bureau of Forest to loosen the qualification for contracting afforestion project and to allow aborigine to participate with the reforest activity increasing their income and resolving living problem.

In this activity, first team of spontaneity named Mess reforest group was established in Nan-Tou Jenia county to propagate the sense of afforest and forest protection, advising landowner put more efforts on reforesting. The goal of the first year was to finish 50 acres.

The team of aboriginal afforest is mainly formed by local old man of the tribe. The reasons to group old man are to provide opportunity of job for them and to use their knowledge on forest ecology due to the long-term living in the mountain area. Through the activity of land recovering and wooding the target of afforestation could be reach much easier.

Result

China Time originated the River Protection Association to execute the river protection joining the social groups in 1990. After 1992, River Protection Association participated in saving water resources and linked up China Time Education Association, Environment Protect Agency, Education Department, Agriculture Department, Economy Department, Forest Department, Chinese daily Newspaper, Chin-Zhi Accounting Company and De-Liu Lawyer Company etc. to plant one hundred dollars NT per tree. For six years, there were two hundred thousands people contributing money and excess five hundred thousands people joined these activities that combined public and government to afforest.

Afforestation must be a continuous activity neither a ceremony nor a tool for election. We known every trees have to care more than six years that the sapling grown up in good condition. For above reason, the activity of "Afforestation saving water Resources" began on 1992. 3 and then we plant every 3 month until now are ten years. We wish these activities either waking people to care the water resources and executing to plant or promoting the concept of public afforestation.

Reference

1. Statistic databases of Environment Protect Agency.
2. Water Resources databases of Economy Department.
3. Statistic databases of Agriculture Department, "The effects of mountain slope planting betel palm".
4. Twenty million trees saving Taiwan water resources Association, "Green earth and eternal river".

An Inquiry System for Design Discharge in Small and Midsize Watersheds

Kwan Tun Lee[1], Chung-Chieh Meng[2], Shen Chiang[2] and Yi-Ru Chung[3]

Abstract

The paper presents a geomorphic and hydrologic information system for design discharge in small and midsize ungauged watersheds. A compound module, which is based on digital elevation model, frequency analysis theory, kinematic-wave approximation, and geomorphic based runoff modeling can provide enough information for design discharge for different return period conditions. Since the location of future hydraulic work are unpredictable, geomorphic information of any location within the watershed is required. While using the inquiry system, users can interactively select a point only by clicking on mouse at the desired location, and then consequent calculation procedures will be conducted. The system also presents the possibility for users to run several location of subwatershed outlet to consult different aspects for alternative designs.

***Keywords**:* design discharge, inquiry system, geographic information system

Introduction

The geographic information system (GIS) has significantly changed the way in acquiring and using spatial data, and provided the potential for the use of physical based watershed hydrologic modeling. Attempts to incorporate GIS into hydrologic analysis can be grouped into four categories (Greene and Cruise, 1995): (1) calculation of input parameters for existing hydrologic models; (2) mapping and display of hydrologic variables; (3) watershed surface representation; and (4) identification of hydrologic response units. Several attempts have been made the GIS techniques applied for watershed hydrologic analysis such as Djokic and Maidment (1991), Greene and Cruise (1995) for urban watersheds; Garrote and Bras (1995) as well as Levy and Baecher (1999) for midsize or large watersheds.

Due to the diverse training of the practical engineers, we found that an inquiry system with highly intuitive and easily operated characteristics is required. This paper presents the development of a system which incorporates the spatial analysis capabilities of the GIS

[1] Prof. and Chairman, Dept. River & Harbor Engrg., Nat'l Taiwan Ocean Univ., Keelung, Taiwan 202, R.O.C. (Email: ktlee@mail.ntou.edu.tw; Fax: 886-2-24634122)

[2] Former Graduate Student, Dept. River & Harbor Engrg., Nat'l Taiwan Ocean Univ., Keelung, Taiwan 202, R.O.C.

[3] Graduate Student, Dept. River & Harbor Engrg., Nat'l Taiwan Ocean Univ., Keelung, Taiwan 202, R.O.C.

characteristics were coded into separate layers, and the attribute information for each layer was used to construct database attribute tables. In contrast to other GIS applications for hydrologic analysis, the coordinate values that define the location of features in the database are used directly to include the spatial heterogeneity of watershed characteristics in the modeling process. The design discharge is based on the kinematic-wave approximation and geomorphic IUH model (Rodriguez-Iturbe and Valdes, 1979). The spatial data sets are used directly in the hydrologic model to determine the design discharge for different return periods.

Watershed Geomorphic Factors Calculation

Channel network extraction from digital elevation data in this study followed the algorithms developed by Jenson and Domingue (1988). The algorithms were performed on a spatial grid system (raster data structure). These techniques are based on neighborhood operations, where calculations and decisions are made for a cell according to the elevations in the eight cells that are spatially adjacent in the raster. The Horton-Strahler ordering system was adopted in this study. The channel network can be extracted using an adequate threshold-area value, and then the length and slope for different order of overland areas and streams as well as other geomorphic factors can be directly obtained from the data set generated by a digital elevation model (DEM). The detail procedure for geomorphic factors calculation can be found in Lee (1998).

Design Discharge Calculation in Small Watersheds

For small watersheds, storm duration usually exceeds watershed time of concentration; that is, the flow equilibrium state is usually reached. The magnitude of watershed peak discharge at equilibrium can be expressed as

$$Q = i_e A \tag{1}$$

in which Q= peak discharge, i_e = rainfall excess intensity, A= watershed area. Once time of concentration is determined, the rainfall intensity can be determined from an intensity-duration-frequency curve storm using rainfall duration equal to the time of concentration. In this study, the estimation of the time of concentration is based on kinematic-wave theory. The application of the kinematic-wave approximation for watershed runoff simulation has been considered acceptable for small watersheds provided the watershed slope is not too flat (Ponce, 1989; Fread, 1993).

While a watershed consisting of two identical rectangular overland-flow planes as a V-shape model, the travel time for the *i*th-order overland-flow is (Henderson and Wooding, 1964)

$$T_{x_{oi}} = \left(\frac{n_o \overline{L}_{o_i}}{\overline{S}_{o_i}^{1/2} i_e^{m-1}} \right)^{\frac{1}{m}} \tag{2}$$

in which n_o is the roughness coefficient for overland planes; $\overline{S}_{o_i}$ is the mean *i*th-order overland slopes; $\overline{L}_{o_i}$ is the mean *i*th-order overland-flow lengths; m is an exponent (=5/3); i_e is the rainfall excess intensity. The travel time for the *i*th-order channel is (Lee and Yen, 1997)

$$T_{x_i} = \frac{B_i}{2i_e \bar{L}_{o_i}} \left[\left(h_{co_i}^m + \frac{2i_e n_c \bar{L}_{o_i} \bar{L}_{c_i}}{\bar{S}_{c_i}^{1/2} B_i} \right)^{\frac{1}{m}} - h_{co_i} \right] \tag{3}$$

in which n_c is the roughness coefficient for channels; $\bar{S}_{c_i}$ is the mean ith-order channel slope; $\bar{L}_{c_i}$ is the mean ith-order channel-flow length; B_i is the width of the ith-order channel; and h_{co_i} is the inflow depth of the ith-order channel due to water transported from upstream reaches. The value of h_{co_i} is equal to zero for i=1 because no channel flow is transported from upstream. For $1 < i \leq \Omega$, h_{co_i} can be expressed as (Lee and Yen, 1997)

$$h_{co_i} = \left[i_e n_c \left(N_i \bar{A}_i - AP_{OA_i} \right) / (N_i B_i \bar{S}_{c_i}^{1/2}) \right]^{\frac{1}{m}} \tag{4}$$

in which N_i = the number of ith-order channels, and $\bar{A}_i$ = the mean of the drainage area of order i. By means of Eqs. 2 and 3, the travel time for different order subwatersheds can be estimated from overland and channel hydraulics instead of relying on watershed empirical formulas.

The time of concentration is the time required for the runoff from the hydraulic remote point to the watershed outlet. For an Ωth-order watershed, the time of concentration is the summation of runoff travel time in overland-flow and consequent channel-flow states, which is

$$T_c = T_{x_{o1}} + \sum_{i=1}^{\Omega} T_{x_i}$$

$$= \left(\frac{n_o \bar{L}_{o_1}}{\bar{S}_{o_1}^{1/2} i_e^{m-1}} \right)^{\frac{1}{m}} + \sum_{i=1}^{\Omega} \frac{B_i}{2i_e \bar{L}_{o_i}} \left[\left(h_{co_i}^m + \frac{2i_e n_c \bar{L}_{o_i} \bar{L}_{c_i}}{\bar{S}_{c_i}^{1/2} B_i} \right)^{\frac{1}{m}} - h_{co_i} \right] \tag{5}$$

in which T_c is the time of concentration of the watershed. The benefits of adopting Eq. 5 for watershed time of concentration estimation is the equation can be applied to watersheds of different geomorphic conditions, and the equation can account for the influence of different rainfall intensities.

Design Discharge Calculation in Midsize Watersheds

For midsize watersheds, the influence of temporal rainfall variation on runoff characteristics is significant. The average rainfall excess intensity in Eq. 1 cannot well generate the hydrologic response of the watershed. For design purpose, the results of the intensity-duration-frequency analysis combining with the alternating block method (Chow et al., 1988) was adopted to derive the design hyetograph in this study, and then a hydrograph routing technique was applied to reflect the time-varying hyetograph on runoff hydrograph.

Conventional unit hydrograph method (Sherman, 1932) or instantaneous unit hydrograph method (Clark, 1945; Nash, 1957; Dooge, 1956) can provide acceptable results for runoff simulation if good quality of hydrologic records is available. For watershed with limited length of flow data or even without flow record, conventional methods will not provide useful helps. In this circumstance, a geomorphic-based runoff model was adopted in this study for runoff simulation in midsize watersheds.

In the geomorphic IUH model, the hydrologic response of the watershed can be expressed as (Rodriguez-Iturbe and Valdes, 1979)

$$u(t)=\sum_{w\in W}\left[f_{x_{oi}}(t)*f_{x_i}(t)*f_{x_j}(t)*\cdots*f_{x_\Omega}(t)\right]_w\cdot P(w) \tag{6}$$

in which w is a specified runoff path, $w \in W$; W is the path space; $*$ denotes a convolution integral; $f_{x_{oi}}(t)$ is the overland-flow travel-time probability-density function in state x_{oi} with a mean value of $T_{x_{oi}}$; $f_{x_j}(t)$ is the channel-flow travel-time probability-density function in state x_j with a mean value of T_{x_j}; $P(w)$ is the probability of a drop of rainfall excess adopting this specified path. The travel time for the overland-flow region can be estimated by using Eq. 2, and the travel time for the channel state can be calculated by Eq. 3. The travel-time probability-density functions in Eq. 6 have been proposed by previous researchers as exponential distributions or uniform distributions (Rodriguez-Iturbe and Valdes, 1979; Gupta et al., 1980). In this study, the travel times for the overland-flow and channel-flow were assumed to follow an exponential distribution. Further detail of the kinematic-wave based GIUH model can be found in Lee and Yen (1997).

Inquiry System and User Interface

From the user's point of view, the inquiry system should provide a convenient figure interface for design work. In order to make the system useful as a decision-support tool, its implementation must allow users for various design conditions. Through the interface, the user can easily access watershed database and model output for specified design conditions. In this study, the interface code was written in Borland C++ language, following an object-oriented design methodology (Cox, 1986), which is predominantly in most of geographic information systems. The GIS functions were used to assign all attributes to corresponding location of the watershed. The strategy in this system to deal with abundant watershed information is to let the user select preferred scale of figures, which are interactively managed by the user. Following the analytical procedure mentioned previously, the core of the inquiry system should be composed of data acquisition, geomorphic factors calculation module, and design discharge calculation module. Each module consists of several executable Fortran programs that operated in the same database environment.

The inquiry system can be run in a PC 486 or higher environment. The user interface module can extract high-level descriptions of watershed state from the database. The inquiry system is flexible to offer detailed information at any point within the study watershed for users' request. Once the user specified the location of subwatershed outlet, abundant calculation work is conducted to delineate the subwatershed boundary out, and then to generate geomorphic factors for the time of concentration calculation. Since the kinematic-wave time of concentration varies with rainfall intensity, users need to specify a design return period, which corresponds with a prepared intensity-duration regression equation.

Case Study

The inquiry system is flexibly applied to any watershed if geomorphic and hydrologic databases are provided for the input requirement of the system. In this study, the Heng-Chi watershed located at northern Taiwan is used as an example to illustrate the capability of the inquiry system. The size of the entire watershed is 53 km^2, and the stream network of the watershed is fourth-order. Recently, rapid developments were found in the downstream part of the Heng-Chi watershed, to avoid serious inundation and sediment deposition in the downstream area, upstream soil conservation and downstream channel regulation works are required. Consequently, it is considered to be useful to establish a convenient inquiry system, which can provide detailed geomorphic and hydrologic information for design work at any desired location within the watershed.

Results and Discussion

The resolution of the digital elevation data set we obtained is $40m \times 40m$. By applying DEM technique for depression elimination, the stream network is extracted from the data set, and then the geomorphic factors can be calculated. Figure 1 is the frontpage of the inquiry system, which shows the watershed map and channel network. The geomorphic factors for the entire Heng-Chi watershed are listed in Table 1. As mentioned previously, the inquiry system can provide geomorphic information not only for the watershed outlet, but also for any location within the watershed. While user clicks on mouse at the desire location (Fig.2), consequent calculation module will be conducted, and then the interface automatically delineates the subwatershed boundary on screen (Fig. 3).

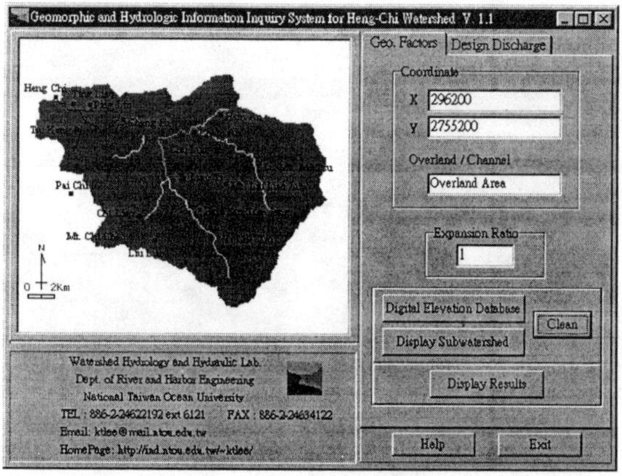

Fig. 1 Heng-Chi watershed map and stream network

Table 1 Geomorphic factors of the entire Heng-Chi watershed

Order i	N_i	$\bar{A}_i$ (km^2)	P_{OA_i}	$\bar{L}_{C_i}$ (km)	$\bar{S}_{C_i}$	$\bar{L}_{O_i}$ (km)	$\bar{S}_{O_i}$
1	30	1.043	0.635	0.659	0.0873	0.854	0.4498
2	6	6.919	0.215	2.743	0.0502	0.348	0.4188
3	2	19.898	0.092	1.601	0.0121	0.764	0.3485
4	1	53.227	0.058	4.970	0.0118	0.310	0.3474

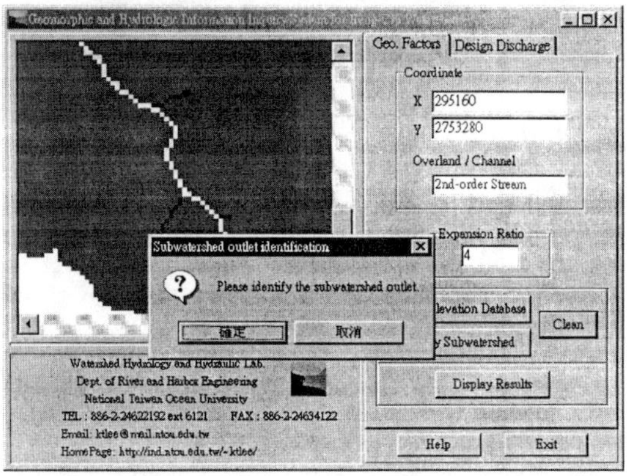

Fig. 2 User specifies a subwatershed outlet

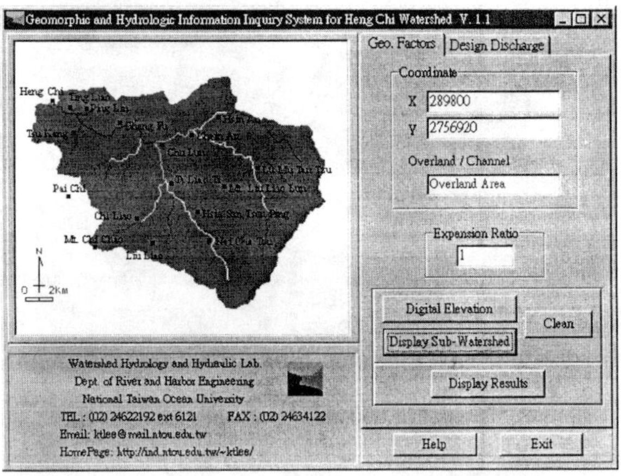

Fig. 3 Delineation of the specified subwatershed

In applying Eqs. 2 and 3, the channel roughness coefficient n_c and overland roughness coefficient n_o need to be determined. Adequate channel roughness coefficients and overland roughness coefficient can be referred to Chow (1959) for channels and HEC (1990) for overland. Once the subwatershed is delineated, the percentage of different land covers within the subwatershed can be calculated based on the land cover database to obtain a corresponding overland roughness coefficient n_o.

The Ta-Pao station is the only available rainfall station in the Heng-Chi watershed. A three-parameter rainfall intensity-duration regression equation was used to represent the rainfall characteristics. The analytical results for different return periods are listed in Table 2. As shown in Fig. 4, user needs to specify a return period, and then the design rainfall intensity i is determined by using the intensity-duration equation combining with the time of concentration equation. The design rainfall-excess intensity i_e is then obtained by deducing 1 *mm/hr* for rainfall abstraction, which is usually adopted in the Taiwan Water Conservancy Institute.

Table 2 Regression coefficients of the intensity-duration-frequency analysis

Return period	5 yr	10 yr	25 yr	50 yr	100 yr	200 yr
a*	834.589	889.882	948.345	990.049	1022.119	1021.746
b	13	14	16	18	20	21
c	0.5786	0.5653	0.5496	0.5394	0.5289	0.5147

* in which $i = a/(t_d + b)^c$; i : rainfall intensity in *mm/hr*, t_d: rainfall duration in *min.*

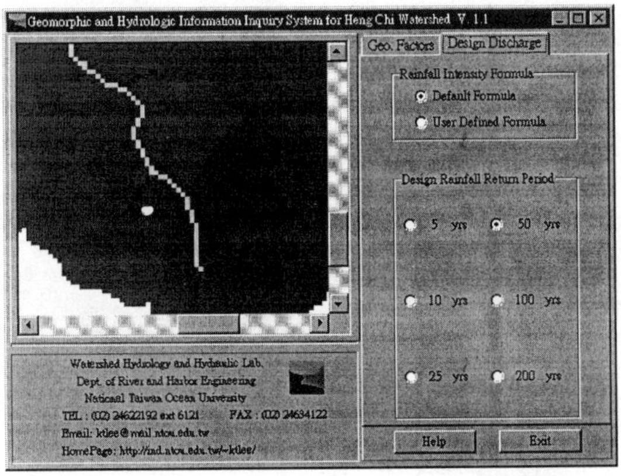

Fig. 4 User specifies a design return period

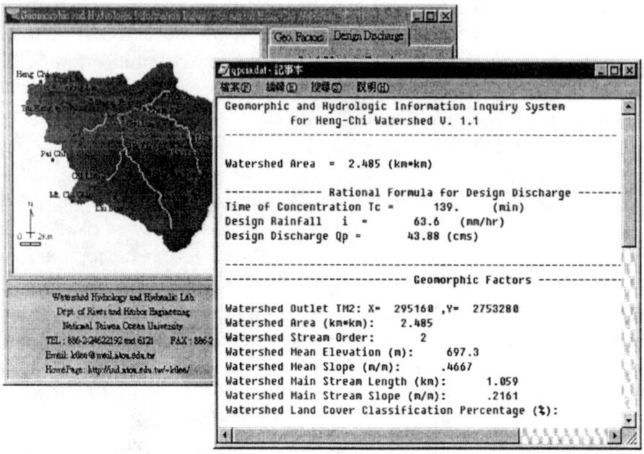

Fig. 5 Geomorphic and hydrologic information output for a small subwatershed

While the area of the specified subwatershed is less than 2.5 km^2, the design discharge is estimated by using Eq. 1. For subwatershed area is larger than 2.5 km^2, the kinematic-wave based GIUH model (Eq. 6) is used to obtain the runoff hydrograph. As shown in Fig. 5, the geomorphic information and corresponding design discharge for a specified subwatershed area less than 2.5 km^2 are shown on the windows. Fig. 6 shows the output of the inquiry system for a subwatershed larger than 2.5 km^2. The user-driven interface can easily access the results and store the information in the database. Moreover, the user can run several locations of subwatershed outlet to consult different aspects and possibility for alternative designs.

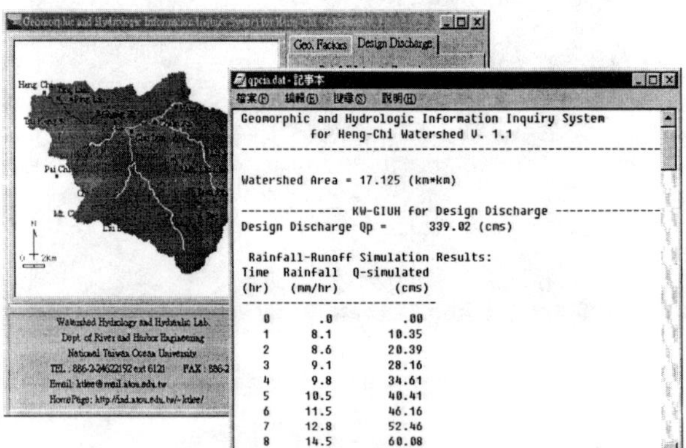

Fig. 6 Geomorphic and hydrologic information output for a midsize subwatershed

Conclusions

In this paper, we present a windows-based inquiry system for watershed design discharge. The user-friendly interface has shown the possibility to implement design work for any location within a midsize watershed. The system integrates a geomorphic and hydrologic database within a graphic computer environment. The computational modules are implemented using object-oriented design techniques; therefore, the modules and database can be used for further various purposes.

Acknowledgments

This study is a research work supported by the National Science Council, Taiwan, R.O.C., under grants NSC 88-TPC-E-019-006 and NSC-89-TPC-7-019-011. Financial support from the National Science Council is gratefully acknowledged.

References

Chow, V. T. (1959). *Open-Channel Hydraulics*, McGraw-Hill, New York, N. Y.

Chow, Maidment, D. R., and Mays, L. W. (1988). *Applied Hydrology*, McGraw-Hill, Inc., New York.

Clark, C. O. (1945). "Storage and the unit hydrograph," *ASCE Trans*, 110, 1419-1446.

Cox, B. J. (1986). *Object-oriented programming: an evolutionary approach*, Addison Wesley, M.A.

Djokic, D. and Maidment, D. R. (1991). "Terrain analysis for stormwater modeling," *Hydrological Processes*, 5(1), 115-124.

Dooge, J. C. I. (1956). "A general theory of unit hydrograph," *J. Geophys. Res.*, 64, 241-256.

Fread, D. L. (1993). "Chapter 10: Flooding routing," *Handbook of Hydrology*, D. R. Maidment, ed., McGraw-Hill Book Co., New York, N. Y.

Garrote, L. and Bras, R. L. (1995). "An integrated software environment for real-time use of a distributed hydrologic model," *J. Hydrol.*, 167: 307-326.

Greene, R. G. and Cruise, J. F. (1995). "Urban watershed modeling using geographic information system," *J. Water Resour. Planning and Management, ASCE*, 121(4), 318-325.

Henderson, F. M., and Wooding, R. A. (1964). "Overland flow and groundwater flow from a steady rainfall of finite duration," *J. Geophys. Res.*, 69(8), 1531-1540.

Hydrologic Engineering Center (1990). *HEC-1 flood hydrograph package: user's manual*, U. S. Army Corps of Engineers, Davis, California.

Jenson, S. K., and Domingue, J. O. (1988). "Extracting topographic structure from digital elevation data for geographic information system analysis," *Photogramm. Eng. Remote Sens.*, 54, 1593-1600.

Lee, K. T. and Yen, B. C. (1997). "Geomorphology and kinematic-wave based hydrograph derivation," *J. Hydraulic Engrg., ASCE*, 123(1), 73-80.

Lee, K. T. (1998). "Generating design hydrographs by DEM assisted geomorphic runoff simulation: a case study," *J. Am. Water Resour. Assoc.*, 34(2), 375-384.

Levy, B. S. and Baecher, G. B. (1999). "Nilesim: a windows-based hydrologic simulator of the Nile River basin," *J. Water Resour. Planning and Management, ASCE*, 125(2), 100-106.

Nash, J. E. (1957). "The form of the instantaneous unit hydrograph," IASH Pub. No. 45, Vol. 3.

Ponce, V. M. (1989). "Section 4.2, Overland flow" and "Chapter 9, Stream channel routing," *Engineering Hydrology*, Prentice Hall Book Co., Englewood Cliffs, N. J.

Rodriguez-Iturbe, I. and Valdes, J. B. (1979). "The geomorphologic structure of hydrologic response," *Water Resour. Res.*, 15(6), 1409-1420.

Sherman, L. k. (1932). "Stream-flow from rainfall by the unit–graph method," *Eng. News-Rec.*, 108, 501-505.

Wet-weather Treatment Process Simulation Using SWMM

Wayne C. Huber[1], M. ASCE

Abstract

The paper summarizes the ability of SWMM to simulate control options and wet-weather controls (WWCs) that are favored in current practice, including those related to low impact development (LID). The model is able to simulate some practices well, such as storage, and other options not as well, such as wetlands and filtration.

Introduction

The U.S. Environmental Protection Agency's Storm Water Management Model (SWMM) dates from 1971 (Metcalf and Eddy et al., 1971) and is currently at Version 4.4h (Huber and Dickinson, 1988; Roesner et al., 1988). A summary of model capabilities is provided by Huber (1995). The most current version of the model may be downloaded from the Oregon State University SWMM web site: www.ccee.orst.edu/~swmm. This web site includes links to several graphical user interfaces (GUIs) that make the model much easier to use. The model is dynamic, and updates to the code occur several times during a year. A SWMM discussion group is available on the Internet through the courtesy of Prof. William James at the University of Guelph in Ontario, Canada. To join this free group, send a message to: list-serve@uoguelph.ca. The text of the message should say only: subscribe swmm-users <your name>. That is, your name should follow the "subscribe swmm-users" words. After joining, any message sent to swmm-users@uoguelph.ca will be sent to all members of the group.

SWMM has enjoyed many modifications in recent years, including several that enhance the model's ability to simulate best management practices (BMPs) and general control options for management of stormwater and combined sewers (Huber, 1996; Huber, 2001). Modifications involving quality simulation include:

- Ability to route overland flow from impervious to pervious surfaces, and vice versa.
- Ability to route overland flow from one subcatchment to another.
- Removal functions for simulation of load removal in Runoff Block subcatchments and Transport Block channel/pipes.

---

1. Department of Civil, Construction, and Environmental Engineering, Oregon State University, Corvallis, Oregon 97331 USA

- Removal by settling in Transport Block channel/pipes.
- Continuity checks to reflect various removal pathways, including by infiltration.

Additional options exist within several SWMM blocks to simulate hydraulic controls, such as time-dependent regulators, weir and orifice settings that depend upon stages in specified locations, etc. These hydraulic controls are located primarily within the Extran Block, which does not simulate water quality. Storage options can be simulated in the Runoff, Transport, Extran (quantity only), and Storage/Treatment Blocks, with the most sophisticated quality options residing in the latter. A recent addition to the Transport Block also provides for simplified modeling of linked DO-BOD-NOD processes, essentially using a modified Streeter-Phelps analysis. This recent (2001) addition also permits removal in Transport Block channels and pipes based on first-order decay, settling (using a constant settling velocity), and a removal fraction (applied to incoming loads). This paper updates the capabilities described by Huber (1996) and provides additional examples.

Wet-Weather Controls

So-called "best management practices" (BMPs) are also called "sustainable urban development systems" (SUDS) in the United Kingdom and may or may not be "best" or "sustainable" in reality. Another term that has entered the jargon recently is "low impact development" or LID, in which an attempt is made to implement water quantity controls at the parcel (local home or small development) level (Prince Georges County, 2000). Again, the impact from LID may or may not actually be "low." Hence, an alternative, less declarative term, is "wet-weather control" or WWC, meaning a control designed to affect water quantity and/or quality during wet weather (Heaney, 2001).

Wet-Weather Control Data

The most complete summary of information on WWC performance is the EPA/ASCE BMP Database project managed by the American Society of Civil Engineers that has received major funding during the past three years from U.S. Environmental Protection Agency (Wright Water Engineers, 2000). The EPA/ASCE BMP database project has been collecting information on a wide variety of WWCs, their performance, and costs. The results, and software to query the database, can be downloaded at http://www.bmpdatabase.org/docs.html. A summary of the attributes in the EPA/ASCE BMP Database is shown in Table 1 (Heaney, 2001). A total of 22 categories of information is available. The first nine entries provide general site information. Entries 10 to 12 are the storm event precipitation, quantity and quality data. The final ten entries are the ten categories of WWCs. The attributes of each category can be partitioned into site, storm event, and cost. About two-thirds of the attributes relate to the site, and 102 of the 359 attributes relate to cost. Storm event attributes for water quality depend upon the number of water quality constituents that are measured.

Table 1. Summary of categories and attributes of entries in EPA/ASCE BMP Database.

Number	Category	Site Attributes	Event Attributes	Cost Attributes	Total
1	General test site information	18	0	0	18
2	Sponsoring and testing agencies	2	0	0	2
3	Watershed information	19	0	0	19
4	Roads and parking lots	16	0	0	16
5	Non-structural BMP data entry	8	0	2	10
6	Structural BMP information	8	0	0	8
7	Monitoring stations	5	0	0	5
8	Site monitoring information	6	0	0	6
9	Site monitoring costs	2	0	9	11
10	Precipitation data	7	0	0	7
11	Storm runoff and base flow data	2	10	0	12
12	Water quality sampling event data	2	12	0	14
13	Detention basin design data	15	0	8	23
14	Grass filter strip design data	10	0	8	18
15	Infiltration basin design data	16	0	9	25
16	Media filter design data	12	0	8	20
17	Porous pavement design data	16	0	10	26
18	Retention pond design data	15	0	12	27
19	Percolation trench and dry well design data	16	0	10	26
20	Wetland channel and swale design data	12	0	8	20
21	Wetland basin design data	20	0	10	30
22	Hydrodynamic devices	8	0	8	16
	Total	**235**	**22**	**102**	**359**

Source data from Wright Water Engineers (2000).

Other sources and potential sources of WWC data include a $30 million effort by the California Department of Transportation (Caltrans) in which a total of 39 structural WWCs are being installed at 33 transportation sites in California. Over two years of monitoring have taken place for these WWCs. Details of the Caltrans programs can be found at:
http://www.dot.ca.gov/hq/env/stormwater/ongoing/bmp_retrofit/bottom2.htm

The first significant results from the Caltrans studies are due in November 2001. Other data sources include studies in Europe (e.g., Chocat, 2001), Asia and Australia. It is imperative to use high-quality data to improve modeling algorithms.

Wet-Weather Control Simulation Options

Among several options, SWMM is able to simulate the effects of storage, I/I (infiltration/inflow) reduction, street cleaning, catchbasin (gully pot) cleaning, high-rate treatment, and many different hydraulic configurations (abetted by the Extran Block and its dynamic wave modeling capability). Improved infiltration options (into the soil) allow better simulation of infiltration-type controls and linkage with shallow groundwater. Overland flow quantity and quality may drain from impervious areas over pervious areas (and vice versa), to permit direct simulation of overland flow infiltration devices and vegetated buffer strips. A list of WWC options and SWMM suitability is given in Table 2.

It is evident from Table 2 that in terms of simulation of quality control options, the model is best suited toward those depending on some form of storage and those relying on reduced surface loadings. Within the Storage/Treatment (S/T) Block, storage-routing itself is easily accomplished, so the model is able to perform required hydraulic computations. Quality removal then depends upon the ability to simulate a process as first-order, for which simple finite-difference formulations exist, or by two more complex mechanisms. (First-order removal is also an option in the Transport Block, assuming complete mixing within storage units.) A second S/T Block mechanism consists of a generalized removal equation, designed such that most empirical or other results can be mimicked. For example, the common exponential removal equation,

$$R = R_{max}(1 - e^{-kt_d}) \qquad (1)$$

can easily be simulated because detention time, t_d, is computed under the assumption of either complete mixing or of plug flow. In Equation 1, R is removal fraction, R_{max} is maximum removal fraction, and k is a first-order decay coefficient. Removal of a given pollutant can be computed as a function of concentration of any pollutant, removal fraction of any other pollutant, and/or detention time. This formulation has proved robust in its ability to reproduce monitored performance characteristics, including such high-rate devices as swirl concentrators and dissolved air flotation. Removal equations are the only way in which non-storage devices can effect removal in the model, such as screens and bar racks.

The third removal mechanism in the S/T Block is by sedimentation, using Camp's (1946) theory for quiescent conditions, modified for turbulence as described by Chen (1975). For this formulation, up to five settling velocity ranges can be input, or alternatively, the five ranges can be computed through input of five specific gravity - particle size ranges. Concentrations within each range are then routed through one or more units, such that if large material is removed in an upstream process, it is no longer available for removal in a downstream process. That is, moving downstream through a progression of (up to five) S/T units, the characteristics of a pollutant change from coarser to finer, assuming large material is removed first (as in a grit chamber).

Table 2. Wet-weather controls and SWMM suitability

WWC Option	SWMM Suitability
Storage and Associated Treatment	Simulation of most storage options, with several options for hydraulic controls. Treatment by removal equations, first-order decay, or sedimentation. Removal in Transport Block channel/pipes by first-order decay or sedimentation (constant settling velocity) or removal fraction applied to incoming loads.
Screening and Filtration	Simulation by removal equations.
Chemical Treatment, Chlorination	Simulation by removal equations.
Wetlands	Simulation only to the extent that wetland behaves like a storage device.
Source Controls	Simulated by reduced loadings.
Street Cleaning	Simulated directly through removal fractions or by reduced loadings.
Overland Flow, Swales, Infiltration, Porous Pavement	Simulation in Runoff Block only. Optional quality removal by first order decay or removal fraction applied to incoming load. Infiltrating water carries pollutants with it.
Maintenance, e.g., Sewer Flushing	Simulation by reduced loading, with option for sediment scour/deposition in Transport Block.
Illicit Connection Removal	Simulation through modification of drainage system connectivity.
Inlet Constrictions	Hydraulic control, simulated in Extran Block.
Real-Time Control	Orifice and weir settings in Extran can be controlled as function of time and/or head changes.

Use of removal equations in SWMM can lead to removal being a function of the initial (incoming) concentration. For instance, monitoring data are sometimes interpreted in the form:

Removal efficiency = 1 – effluent concentration/influent concentration (2)

But observe in this equation, that removal efficiency will increase whenever influent concentration increases. Thus, if effluent concentrations are somewhat constant, artificially high removal efficiencies may result simply because of high influent concentrations. Indeed, in an analysis of the EPA/ASCE BMP Database, Strecker et al. (2001) observed that WWC percent removal increases with increasing influent concentration. They postulate that several WWCs can be characterized simply by the effluent concentration distribution, in lieu of a heuristic or process-model of the system. While the ASCE database data appear strongly to lead to this conclusion, to date the data are not sufficient to tie designs (WWC sizing, processes) to effluent quality. In addition, effluent quality would be difficult to use to characterize WWCs in series (a "treatment train"). Hence, process models, of the nature of SWMM, are useful for analysis of WWC performance.

An improvement to Equation 1 might be of the form:

$$\frac{C(t)-C_o}{C_i-C_o}=e^{-kt_d} \qquad (3)$$

where C_i = influent concentration and C_o = effluent concentration, characterized from data such as in the EPA/ASCE BMP Database. Then the initial concentration "decays" not to zero but to a value representative of monitoring data for the device. The C_o parameter could also vary (e.g., lognormally) over the range of observed observations. This feature is not yet implemented in SWMM.

Most other WWC options affecting quality are performed in the SWMM Runoff Block, in which rainfall is converted to runoff and initial surface loadings are generated. Loadings are computed by a variety of mechanisms, including buildup-washoff, rating curves (regression of load vs. flow), constant concentrations (e.g., use of an event mean concentration or EMC), catchbasin (gully pot) flushing, precipitation loads, and land-surface erosion. The only land-surface removal process explicitly incorporated into the model is street cleaning. This has sometimes led to optimistic performance predictions for street cleaning, under the assumption that most pollutants are available for removal by modern street sweepers and that such sweepers perform as indicated in carefully controlled tests. While street cleaning may, in fact, reduce pollutant concentrations in runoff, frequent (e.g., daily) cleaning may be required for significant improvement in stormwater runoff quality (Pitt, 1979). On the other hand, more recent studies indicate that technological advances can produce much better results (Sutherland and Jelen, 1997).

Source controls can also be simulated by reducing loading rates under the buildup-washoff mechanism. But this is strictly a heuristic procedure, devoid of physical mechanisms. For example, if studies of public awareness campaigns illustrate a load reduction of 15%, then loading rates can be reduced by this amount. Unfortunately, there are few such data to support such claims, however.

Quality loads can also be contributed by direct links to the sewer system from industries and businesses. Limited flow and quality data from business do exist, as documented for the SWMM Transport Block (e.g., Metcalf and Eddy et al., 1971; Huber and Dickinson, 1988). Illicit and improper connections to the sewer system can be eliminated in the model and the impacts of such disconnections studied easily.

Storage Example

Hypothetical storage is simulated downstream of the 5.95 ha existing high-density residential (King's Creek Apartments) area in Miami, Florida, monitored by the U.S. Geological Survey (Miller, 1979). The hypothetical 1,022 m^2 storage unit of volume 1,493 m^3 is placed at the outlet. Total nitrogen (TN) and ortho-phosphorus (OPO4-P) are removed according to Equation 1, with the parameters shown in Table 3.

Table 3. Exponential removal parameters for TN and OPO4-P.

Parameter	TN	OPO4-P
R_{max}	0.1	0.2
k (day^{-1})	10	10

The parameters are chosen consistent with literature values (Urbonas, 1993). A third pollutant, total suspended solids (TSS) is simulated using sedimentation theory. Full parameter input is not provided because of space limitations. Output from an eight-year simulation, 1984-1991, is shown in Table 4.

With a maximum treatment efficiency of 10% for TN and 20% for OPO4, actual removals are 6.6% and 13.4%, respectively (from the ratio of net inflow to net outflow). These are less than the maximum because sufficient detention time is not always reached. On the other hand, sedimentation removes almost 97% of the TSS, indicating once again the relative ease with which particulates can be removed as opposed to soluble material. The predicted TSS removal may be too high; a review of BMP efficiencies by Urbonas (1993) indicates values on the order of 90% for TSS. Although not shown, a larger basin, with longer detention time, yields higher removal percentages for all pollutants, on the basis of another SWMM simulation.

Table 4. Example summary output from SWMM Storage/Treatment Block.

Parameter	Volume	TN	OPO4-P	TSS
	m^3	kg	kg	kg
Inflow, total	3.43E+05	9.25E+02	4.16E+02	2.48E+05
Inflow, net	3.43E+05	9.25E+02	4.16E+02	2.48E+05
Bypass	0.00E+00	0.00E+00	0.00E+00	0.00E+00
Treated outflow	3.40E+05	8.64E+02	3.61E+02	7.90E+03
Residual flow	0.00E+00	0.00E+00	0.00E+00	0.00E+00
Remaining	4.09E+02	6.15E+01	5.58E+01	2.40E+05
Evaporation	2.41E+03			

Other output of interest shown in Table 3 indicates that there were no bypasses, that is, the entire catchment runoff passed through the detention pond, and that evaporation plays a significant role in the water budget.

Overland Flow Example

In addition to using the S/T Block of EPA SWMM, Huber (2001) shows how the performance of many of the WWCs can be simulated using the Runoff Block of SWMM. The Runoff Block has been modified to permit alternative flow pathways during overland flow that route flow from impervious areas over pervious areas, for instance, as shown in Figure 1. The concept of low-impact development (LID) is to minimize off-site flows as much as possible by encouraging infiltration as near to the runoff source as possible. New "green" stormwater management technologies increasingly emphasize overland flow and infiltration as WWCs. These methods attempt to store and infiltrate as much stormwater runoff on-site as possible, for both residential and larger structures. For instance, infiltration, water reuse, evaporation, and transpiration can be encouraged through practices such as directing roof runoff onto lawns, use of rooftop vegetation and gardens, use of infiltration trenches near buildings, etc. Many of these options may be simulated using the most recent version of SWMM (version 4.4h) distributed on the Oregon State University (OSU) web site given earlier.

These changes can have a major impact on the predicted downstream hydrograph as illustrated in Figure 2 (Huber 2001). Routing the flow from the impervious area to the pervious area reduces the peak discharge from 2.9 cfs to about 0.3 cfs, a 90% reduction. Water quality parameters can also be routed in a similar fashion. There is no limit on the WWC treatment train that can be modeled in this manner.

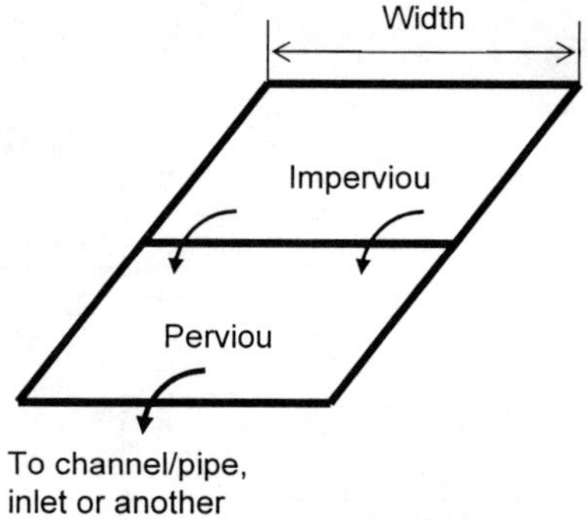

Figure 1. Conceptual routing from the impervious sub-area of a subcatchment to the pervious sub-area of a subcatchment. The scheme is similar for flow from pervious to impervious. Subcatchment outflow can also be directed to another subcatchment.

Overland Flow Volumes and Pollutant Loads

In addition to peak flow reduction, re-direction of flow from impervious to pervious surfaces also can reduce the volume of runoff (as is evident by comparison of the hydrographs in Figure 2) the corresponding pollutant loads. Volumes and masses corresponding to conditions "a" (no redirection of overland flow) and "b" (redirection of impervious runoff to pervious areas) above are shown in Table 5. The reduction in load does not correspond to the reduction in runoff volume because some of the nitrogen load is generated on the pervious surface and thus is not subject to an additional reduction by redirection. However, this could be accomplished by directing the total subcatchment runoff onto another, pervious subcatchment, as suggested in Figure 1.

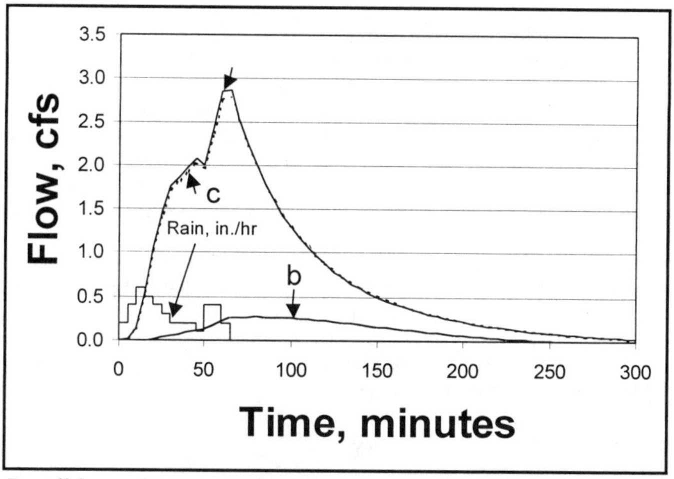

Figure 2. Runoff from subcatchment with 50% imperviousness – a. Traditional Runoff Block overland flow routing; b. Runoff from impervious sub-area flows ove: pervious sub-area; c. Runoff from pervious sub-area flows over impervious sub-area (Huber 2001).

Table 5. Volumes and pollutant load comparison.

Case	Runoff Volume, m^3	NO2+NO3 Load, kg
a. No redirection of overland flow, 50% impervious	400	2.0
b. Redirection of overland flow from impervious area to pervious area	58	0.61
Percent reduction by infiltration	86	70

Summary and Conclusions

SWMM is most effective in simulation storage-type wet-weather controls because such devices have been extensively studied and the most information is available about these widely-used WWCs. Within the Storage/Treatment Block, the model is able to accept most forms of empirical relationships descriptive of treatment efficiency as well as Camp's sedimentation theory. In addition, first-order decay, constant removal fraction (as function of incoming load), and constant settling may be simulated in the Transport Block. A S/T Block example simulation shows that full removal for the situation in which removal is an exponential function of detention time is not achieved unless a larger basin with more detention time is used. Iterative application of a model such as SWMM can be used to design a facility with the desired combination of quantity and quality control.

Overland flow infiltration-based controls may be simulated in the Runoff Block by redirecting runoff from impervious areas onto pervious areas and from one subcatchment onto another subcatchment. This option makes the model more useful for evaluation of control strategies such as "low impact development" that attempt to minimize wet-weather runoff from a development. However, a significant feature lacking from the Runoff or Transport Block is to be able to simulation infiltration from *channels* as well as from overland flow planes. Hence, swales or porous channels in which routing effects may be important may not yet be easily simulated in the model, but this improvement is planned for future releases of the model.

Although not documented in this paper, SWMM also handles hydrologic and hydraulic controls to the extent that they may be included explicitly in one of the model blocks. For example, removal of illicit connections may be simulated in the Transport Block, and hydraulic constrictions and real-time control simulated in the Extran Block. But inferences about water quality may be indirect and based only on reductions of flows, especially for the Extran Block that does not perform water quality simulations.

Acknowledgment

The author acknowledges the assistance of EPA Cooperative Agreement No. CX-826257 in support of portions of this research. The opinions expressed herein do not necessarily reflect those of the Environmental Protection Agency.

References

Camp, T.L. (1946) "Sedimentation and the Design of Settling Tanks," *Transactions*, American Society of Civil Engineers, Vol. 111, pp. 895-936.

Chen, C.N. (1975) "Design of Sediment Retention Basins," *Proc. National Symposium on Urban Hydrology and Sediment Control*, Publication UKY BU109, University of Kentucky,

Lexington, KY, pp. 285-298.

Chocat, B., Barraud, S. and E. Alfakih (2001) "Development of BMPS: Recent Progress in France and Other Countries of Western Europe." *Proc. Linking Stormwater BMP Designs and Performance to Receiving Water Impacts Mitigation*, United Engineering Foundation, Snowmass, CO, August.

Heaney, J.P. (2001) Personal communication, University of Colorado, Boulder, CO.

Huber, W.C. (1995) "EPA Storm Water Management Model -- SWMM," Chapter 22 in *Computer Models of Watershed Hydrology*, V.P. Singh, ed., Water Resources Publications, Highlands Ranch, CO, pp. 783-808.

Huber, W.C. (1996) "BMP Simulation Using the EPA Storm Water Management Model (SWMM)," *Proc. Seventh International Conference on Urban Storm Drainage*, University of Hannover, Hannover, Germany, September, Vol. III, pp. 1629-1633.

Huber, W.C. (2001) "New Options for Overland Flow Routing in SWMM," *Proceedings, World Water and Environmental Resources Conference*, Urban Drainage Modeling Symposium, ASCE, Environmental and Water Resources Institute, Orlando, FL, May 2001, CD-ROM, 8 pp.

Huber, W.C. and Dickinson, R.E. (1988) *Storm Water Management Model , Version 4, User's Manual*, EPA/600/3-88/001a (NTIS PB88-236641), Environmental Protection Agency, Athens, GA.

Metcalf and Eddy, Inc., University of Florida, and Water Resources Engineers, Inc. (1971) *Storm Water Management Model, Volume I -- Final Report*, 11024DOC07/71 (NTIS PB-203289), Environmental Protection Agency, Washington, DC.

Miller, R.A. (1979) *Characteristics of Four Urbanized Basins in South Florida*, U.S. Geological Survey, Open-File Report 79-694, Tallahassee, FL.

Pitt, R. (1979) Demonstration *of Nonpoint Pollution Abatement through Improved Street Cleaning Practices*, EPA-600/2-79-161 (NTIS PB80-108988), Environmental Protection Agency, Cincinnati, OH.

Prince Georges County (2000) *Low-Impact Development Design Manual.* Dept. of Environmental Resources, Prince Georges County, MD.

Roesner, L.A., Aldrich, J.A. and Dickinson, R.E. (1988) *Storm Water Management Model User's Manual Version 4: Extran Addendum*, EPA/600/3-88/001b (NTIS PB88-236658), Environmental Protection Agency, Athens, GA.

Strecker, E.W., Quigley, M.M., Urbonas, B.R., Jones, J.E. and J.K. Clary (2001) Determining Urban Storm Water BMP Effectiveness. *J. Water Resources Planning and Management*, Vol.127, No. 3, pp. 144-149.

Sutherland, R.C. and S.L. Jelen, (1997) "Contrary to Conventional Wisdom, Street Sweeping Can be an Effective BMP," Chapter 9 in *Advances in Modeling the Management of Stormwater Impacts*, W. James, ed., Proc. of the Stormwater and Water Quality Modeling

Conference, Toronto, February 1996, Computational Hydraulics International, Guelph, ON, pp. 179-190.

Urbonas, B. (1993) "Assessment of BMP Use and Technology Today," *Proceedings Sixth International Conference on Urban Storm Drainage*, Niagara Falls, Ontario, Seapoint Publishing, Victoria, BC, Volume I, pp. 927-932.

Wright Water Engineers, Urban Drainage and Flood Control District, and GeoSyntec Consultants (2000) Determining Urban Best Management Practice (BMP) Removal Efficiencies, Task 1.1-National Stormwater BMP Database Elements. Report to Office of Water, U.S. Environmental Protection Agency, Washington, DC. Available at http://www.bmpdatabase.org/docs.html.

The Relationship between Boussinesq Equation and Non-linear Storage of Baseflow Simulation

Ching-pin Tung[1], Nien-ming Hong[2], Chu-hui Chen[3] and Yih-chi Tan[4]

Abstract

There are several approaches for the analysis of baseflow which include different kind of conceptual models and theoretical analysis. The concepts of linear and nonlinear storage are often used in modeling the baseflow and the nonlinear relationship between baseflow (Q) and moisture storage (S) can be described as $S = aQ^b$. Boussinesq equation is used to predict the baseflow, and the results are as good as the estimation by nonlinear storage. The purpose of this paper is to discuss the relationship between nonlinear storage concept and the Boussinesq Equation. The parameters a in nonlinear storage modeling will be discussed. By using the observed streamflow data in Taiwan, the physical properties of parameter a will be identified. The present study concentrated on the structure of baseflow modeling and their parameters.

Introduction

Baseflow plays an important role in water resources management. Specially, baseflow in drought is a stable water supply. Reasonable estimation of the baseflow is indispensable in irrigation, water power plants, and river pollution management. There are two approaches for the baseflow modeling which are statistical approach and physical approach. In statistic models, baseflow is analyzed with watershed characteristics, porosity, and soil properties. For physical approach, baseflow is estimated by ground water theories. On the other hand, linear storage and nonlinear storage are often used in baseflow estimation. The present study concentrated on the derivation of nonlinear model from physical principle.

The purpose of this study is to identify the relationship between the Boussinesq equation and statistical nonlinear storage model. Statistical nonlinear storage model is often used in baseflow estimation with reasonable results (Wittenberg ,1999; Kubota and Sivapalan, 1995). Even the statistical nonlinear model is built from ground water theory. But the transform from baseflow theories exists some defect. Including time disagreement and water

---

[1] Associate Professor, Department of Bioenvironmental Systems Engineering, and Associate Research Fellow, Hydrotech Research Institure, National Taiwan University, Taiwan, R.O. C.

[2] Ph.D. Candidate, Department of Bioenvironmental Systems Engineering, National Taiwan University, Taiwan, R.O. C.

[3] Professor, Department of Civil Engineering, Chung Kuo Institute of Technology, Taiwan R.O. C.

[4] Professor, Department of Bioenvironmental Systems Engineering, and Research Fellow, Hydrotech Research Institure,National Taiwan University, Taiwan, R.O. C.

table time-variance, all of the above mentioned factors will affect the estimation of nonlinear storage model. It is proposed that there is a physical relation between the nonlinear model and ground water theories.

Review

In baseflow analysis, graphical methods have been used to construct a master recession curve (Tobes, 1969). The statistical methods are applied to develop the relationships between the discharge and geo-hydrological factors, such as the first-order regression model by Vogel and Kroll (1991) and auto-regression and moving average (ARMA) model by Spolia and Chander (1974). Watershed properties such as main stream length, watershed wide, porosity, drainage area and hydraulic conductivity are sometimes included in the statistic model. Moreover, region properties, rainfall pattern, vegetation, topography, are also included. It limited the applications of these model, because different region have different values of parameters. Because of the above discussed limitation of the statistical approach, the physical approaches are used more attractive in the baseflow estimation. The well-know Boussinesq's equation is derived directly to estimate the baseflow. Two main boundary conditions, partial penetrating layer and fully penetrating layer, are applied in Boussinesq's equation (Singh, 1968). Streambed above the impermeable layer is named partial penetrating. On the other hand, streambed at impermeable layer is named fully penetrating. Corresponding, linear storage and nonlinear storage concepts are built.

To obtain analytical solution, groundwater governing equations are assumed that aquifer is uniform, isotropic, no loss from evapotranspiration and no inflow from upstream. The Boussinesq equation and its assumptions are most often cited.

$$\frac{\partial h}{\partial t} = \frac{k}{f}\frac{\partial}{\partial x}\left(h\frac{\partial h}{\partial x}\right) = \frac{k}{f}\left[h\frac{\partial^2 h}{\partial x^2} + \left(\frac{\partial h}{\partial x}\right)^2\right] \quad (1)$$

Where k is hydraulic conductivity and f is porosity. Analytical solutions of baseflow can be derived from Boussinesq equation by using these two boundaries. The impermeable layer assumed to be horizontal may be different from real cases, but the effect of the slope of impermeable layer on baseflow is not apparent some days after rainfall (Zecharias and Brutsaert, 1988). In partial penetration, the water table is linearized with small slope and an average depth D_l. For a consine form profile, Boussinesq's equation is modified as

$$\frac{\partial h}{\partial t} = \frac{k}{f} D_l \frac{\partial^2 h}{\partial t^2} \quad (2)$$

D_l is the initial average water table. The analytical solution of eq. (2) is shown as

$$q = 2kpD_l(D_l - D_c)B^{-1} \sum_{n=1,3,..}^{\infty} Exp(-\frac{n^2\pi^2 kpD_l}{4B^2 f}t) \quad (3)$$

where q_t is instant outflow per unit length of stream channel and p is a constant and about 0.3465. The q_t was denoted in lower case to differ from Q_t that represents average or daily total flows in this study. Eq. (3) is a well-known equation for horizontal groundwater outflow. Assumed for a long period of time, eq. (3) can be further simplified as eq. (4).

$$q = 2kpD_l(D_l - D_c)B^{-1} Exp(-\frac{\pi^2 kpD_l}{4B^2 f}t) \quad (4)$$

For fully penetrating condition, Polubarinova-Kochina (1962) assumed that $D_c = 0$ and groundwater table was an incomplete Beta function. The initial height of water table at x = B is D_b, and then baseflow could be expressed as eq. (5).

$$q = 0.862kD_b{}^2 B^{-1}[1 + (1.115kD_b / fB^2)t]^{-2} \quad (5)$$

where D_b is the initial water table height at x = B and is not the average height as D_l in equation (4). Both D_b and D_l have different heights after different rainfall events. In hydrological systems, eq. (4) and eq. (5) are used as a starting point of physical models. The watershed is lumped as a storage element (Brutsaert and Neiber, 1977). A conceptual equation is

$$Q = K \cdot S^n \quad (6)$$

where Q is the outflow of watershed, S is the volume of water storaged, and K and n are constants. For linear model, with n=1 (Dooge, 1973), combined with water balance equation. Eq.(4) is changed as

$$Q = Q_0 \cdot e^{-c_1 t} \quad (7)$$

where Q_0 is the outflow at t = 0 and c_1 is constant. The linear model and baseflow estimation are applied and examined successfully in some studies (Nathan and McMahon, 1999; Wood et al., 1992; Schwarze et al., 1997). For nonlinear model, when $N \neq 1$, one finds in general

$$Q = Q_0 \cdot (1 + c_2 \cdot t)^{n/(1-n)} \quad (8)$$

where c_2 is a constant. Specially, when n=2, the drainage characteristics in eq. (5) is the same that in eq. (6). Boussinesq (1904) has presented the nonlinear model. The nonlinear relationship was pointed by Coutagne (1948), Kubota and Sivapalan (1995), Wittenberg (1994) and Wittenberg (1999). Wittenberg (1999) has shown that storage-discharge relationship is nonlinear with discharge being proportional to the square of groundwater storage in 100 watersheds. The coefficient of storage is determined by interactive least-square method.

However, coefficients c_1 of eq. (7) and c_2 of eq. (8) are not constant. After precipitation recharge groundwater, c_1 and c_2 are dependent on Q_0. So using different historical baseflow data to determine the coefficients K in eq. (6) would be bias. c_1 and c_2 are not a physical parameters. And nonlinear model is a statistic model. But it can't ignore that nonlinear model is still useful tool in baseflow estimation. The present study proposed that there is relation between nonlinear model and baseflow theories. Physical parameters and it's characteristics were hardly excluded from the regression equation of statistical analyses (Wallis, 1965).

The relationship between Boussinesq equation and non-linear storage of baseflow simulation

The time disagreement and relationship are discussed in this section. In baseflow analysis, average daily streamflow are used to describe the properties. Brutsaert and Lopez (1998), Troch et al. (1993), and Wittenberg (1999) analyzed daily streamflow to improve the characteristics of baseflow. In fully penetrating, eq. (5), the discharge is presented as instant streamflow.

1.Daily baseflow in fully penetration

Instant baseflow per river length is presented in equation (5) for fully penetrating condition. Then average flow per stream length from t1 to t1+T can be obtained as equation (9).

$$\bar{q}_t = \frac{0.862 B^3 D_b^{\ 2} f^2 kT}{(B^2 f + 1.115 D_b K t1)(B^2 f + 1.115 D_b K(T + t1))} \tag{9}$$

Eliminating the term of t1 and integration of the main stream length, L, the relationship among $\bar{Q}_t$, $\bar{Q}_{t+T}$ and $\bar{Q}_{t+2T}$ can be as eq. (10).

$$\frac{1}{\bar{Q}_t} - 2\frac{1}{\bar{Q}_{t+T}} + \frac{1}{\bar{Q}_{t+2T}} = \frac{2 \cdot 1.115^2}{0.862} \cdot \frac{k}{2LB^3 f^2} \cdot T^2 \tag{10}$$

where $\bar{Q}_t$ is the average streamflow at t day. In eq. (10), $\frac{2 \cdot 1.115^2}{0.862} \cdot \frac{k}{2LB^3 f^2} \cdot T^2$ is geo-hydrological factor and time-variance without the effects of water table. Parameter C is defined as $\frac{2 \cdot 1.115^2}{0.862} \cdot \frac{k}{2LB^3 f^2} \cdot T^2$. Eq. (10) is also a good estimator of baseflow from $\bar{Q}_{t-T}$ and $\bar{Q}_{t-2T}$.

2. The relationship of nonlinear storage and fully penetration equation

From water balance equation and eq. (11), Wittenberg (1999) showed the baseflow decay with time as eq. (12).

$$S = aQ^b \tag{11}$$

$$Q_t = Q_0 \left[1 + \frac{(1-b) Q_0^{1-b}}{ab} t \right]^{1/(1-b)} \tag{12}$$

When b = 0.5 (n = 2), eq. (10) is expressed by eq. (12). The relationship among $\bar{Q}_t$, $\bar{Q}_{t+T}$ and $\bar{Q}_{t+2T}$ is shown as eq. (13).

$$\frac{1}{\bar{Q}_t} - 2\frac{1}{\bar{Q}_{t+T}} + \frac{1}{\bar{Q}_{t+2T}} = 2\left(\frac{1}{a^2}\right) \tag{13}$$

Therefore, compared $\frac{2 \cdot 1.115^2}{0.862} \cdot \frac{k}{2LB^3 f^2} \cdot T^2$ in eq. (10) and a in eq. (13), it found

$$a = [\,2(\frac{2 \cdot 1.115^2}{0.862} \cdot \frac{k}{2LB^3 f^2} \cdot T^2)]^{-0.5} \tag{14}$$

The characteristics of coefficient a are physical properties, not including average water table height. By comparing eq. (10) and eq. (14), it is found that the term derived from statistical approach have the power of –0.5 with the term derived from the physical principle.

Case Study

The Lao-Nung watershed in the south of Taiwan was selected. Above 85% precipitation is concentrated from May to Oct. So baseflow is an important water resource in this area. The area of watershed is 812 km^2. The records of streamflow are from 1978-1992. In analysis, the data was selected during five continuous dry days for eliminating the effects of interflow, runoff, and river storage. The streamflow in third, forth and fifth day are used to estimate the geo-hydrologic parameter. Because the wide of this watershed is about 11 km, the magnitude of the parameter is very small and heavily influenced by the errors in observed data. Two methods to determinate the parameter are used. One is to average that in each recession segment, C_{mean} ; the other is to choose the mode of the parameter, C_{mode}. Another analyses have performed by using the segment size of 10 days. Because the storage S is changing with time when continuously estimation in 10 days segment. It means that the parameters that derived satisfy the water balance relation implicitly.

Results and discussion

The parameters, C_{mean} and C_{mode}, are calculated. The results show that C_{mean} =-0.0001 and C_{mode} = 0.0004. The mode is built from 29 times in 238 records. A physical quantity C_{mean} is too close to the value of zero which is not a reasonable answer. i.e: watershed is large. This quantity is highly influence by the accuracy of the observation. Based on the above discussion, C_{mode} is selected and used for analysis. The nonlinear model is built as $S = 70.11Q^2$, where the coefficient 70.11 is from the relationship between a and C in eq. (14). There are 17 recession segments after data selecting. From the initial baseflow in each segment, the initial storage is estimated. The relationship between storage and streamflow is presented in Fig. 1. The storage in watershed is from 600mm to 200mm. The results of nonlinear model estimation are show as Fig. 2. The correlation between observed data and estimated one is about 0.992. It is underestimated about 9%. The results seem perfect in baseflow estimation. But there are two different baseflow distributions in Fig. 2. It is separated at 50 CMS. The recession segments should be selected more carefully.

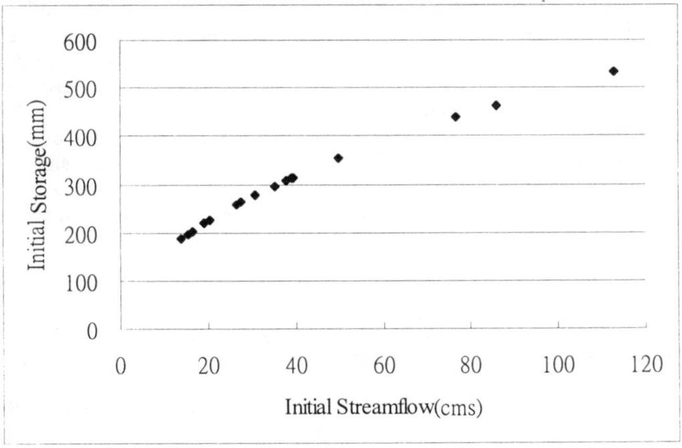

Fig. 1 The initial storage estimation

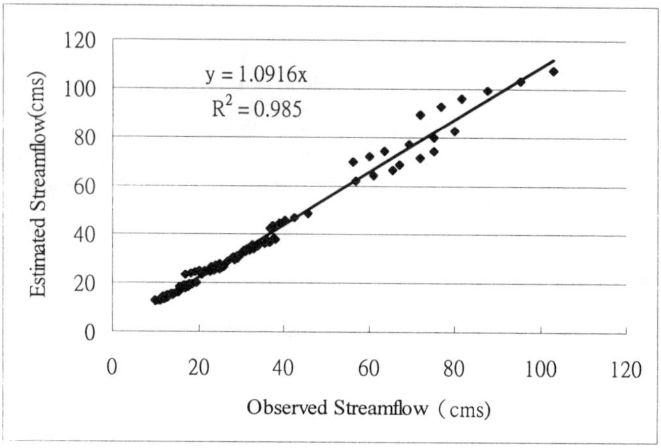

Fig. 2 The baseflow estimation by nonlinear model

Conclusions

Traditionally, the physical water storage phenomenon is described by the use of the statistical model. A new approach is suggested in the present study which uses the Boussinesq equation in the fully penetration condition. The results from the present approach for the continuous baseflow estimation is the same as the result from the nonlinear storage model. These properties suggested that the unknown parameters for the non-linear model can be estimated by the use of geo-hydrological properties. There is a concern for the parameter in the equation of continuous magnitude of the value is small, a little perturbation in the observation data will causes unreliable result.

The present study only use the segment of continuous five dry days of data, different segment size is suggested on the future studies to exanimate the influence on the result. Moreover, in order to fully understand the water storage phenomenon, the effect of rainfall discharge should be included in the future studies. The change of storage will be understand.

References

Boussinesq, J., Recherches theoriques sur l'ecoulenent des nappes d'eau infiltrees dans le sol et sur et sur le debit des sources, J. Math. Pures Appl., 5th Ser., 10, 5-78, 1904.

Brutsaert, W., and J. L. Nieber, Regional drought flow hydrographs from a mature glaciated plateau, Water Resources. Res., 13(3), 637-643, 1977.

Brutsaert, W., and J. P. Lopez, Basin-scale geohydrologic drought flow features of riparian aquifer in the southern Great Plains, Water Resources. Res., 34(2), 223-240, 1998.

Coutagne, A., 1948, "Etude generale des variations de debits en fonction des facteur qui les conditionent, 2eme partie: les variations de debit en periode non influencee par les precipitations", La Houille Blanchu, Sept.-Oct., 416-436.

Dooge, J. C. I. “Linear theory of hydrological systems”, U. S. Dep. Agr. Tech. Bull., 1468, pp.327,1973.

Kubota, J. and Sivapalan, M., “Toward a catchment –scalel model of subsurface runoff generation based on synthesis of small-scale process-based modelling and field studies”, Hydrological Process., 9, 541-554, 1995.

Nathan, R. J. and McMahon, T. A., “Evaluation of automated techniques for base flow and recession analyses”, J. Hydrol., 30, 1465-1473,. 1990.

Polubarinova-Kochina, P. Y. –A., Theory of ground water management, translated from Russian by R. J. M. DeWiest, 613, Pricenton University Press, Preston, N. J. 1962.

Schwarze, R., Droge, W., and Opherden, K., “Regional analysis and modelling of groundwater runoff components from small catchments in solid rock areas”, Landschaftsokol. Umweltforsch., 25, 59-62., 1997.

Singh, K. P., “Some factor affecting baseflow”, 4(5), pp.985-999, 1968.

Spolia, S. K. and Chander, S., Modelling of surface runoff systems by an ARMA model, J. Hydrology, 22, 317-332, 1974.

Toebes, C and Strang, D. D.,” On recession curves, 1-recession equations”, J hydrol. (N.Z.), 3(2), 2-15, 1964..

Troch, P. A., Troch F. De. and W. Brutsaert, Effective water table depth to describe initial conditions prior to storm rainfall in humid regions, Water Resources. Res., 29(2), 427-434, 1993.

Vogel, R. M., and C. N. Kroll, The value of streamflow recession augmentation procedures in low flow and flood flow frequency analysis”, J. Hydrology, 125, 259-276, 1991.

Wallis, J. R., “Multivariate statistical methods in hydrology comparison using data of know functional relationship”, Water Resources. Res., 1(4), pp.447-461, 1965.

Wittenberg, H., “Nonlinear analysis of flow recession curves”, IAHS Publ., 221, 61-67, 1994.

Wittenberg, H., 1999, ”Baseflow recession and recharge as nonlinear storage processes”, Hydrol. Process. 13, 715-726.

Wood, E. F., Letternmaier, D. P., and Zartarian, V. G., “A land-surface hydrological parameterization with subgrid variability for general circulation models”, J. Geophys. Res., 97, 2717-2728, 1992.

TRANSPORT MODELS FOR SOLUTE TRANSFER TO SURFACE RUNOFF

Jiun-Liang Lai[1], Shang-Lien Lo[2], M. ASCE and Shaw L. Yu[3], M. ASCE

Abstract

Two models were discussed to describe the solute transfer to surface runoff. One is a scouring-based model and the other is a diffusion-based model. The scouring-based model is a two-stage model. While the first stage is a quick decrease of mass loss of solute to surface runoff, the second stage is a slow one. The diffusion-based model assumed that solute transfer follows the Fick's law. In this study, a laboratory flume with a medium packed bed was designed. Four different sizes of uniform glass beads were chosen to be the media saturated with three different concentration of potassium chloride before runoff occurred. In a series of experiments, runoff was passed at varied flow rates, velocities, and depths over the medium bed. Runoff samples were taken at the end of flume and analyzed the concentration of potassium chloride. The relationships between the dimensionless modeled parameters and controlled variables were decided by multiple regression. On the other hand, two different soil conditions were selected to be the media. One soil condition was passed 9.5mm screen and the other was passed 4.0mm screen. The same processes of experiment were performed by using these two soil conditions. The outflow concentration of soil experiments was simulated by these two models. The results showed that the simulation of large soil condition was good, but the fine one was underestimated.

Introduction

The nonpoint-source pollution resulting from transport of agricultural chemicals, such as nutrients, pesticides, and other dissolved hazardous chemicals, into surface runoff affects the surface water quality during the period of storm (Miller et al., 1982; Owens et al., 1984; Sharpley et al., 1985; Haith, 1986; Li et al., 1990). Donigian et al. (1997) developed the Agricultural Runoff Management Model (ARM) assuming that the rainfall mixed completely with the soil

[1] Former Graduate Student, Grad. Inst. of Env. Eng., National Taiwan University, Taipei 106, Taiwan
[2] Prof., Grad. Inst. of Env. Eng., National Taiwan University, Taipei 106, Taiwan
[3] Prof., Depart. of Civil Eng. and Appl. Mech., University of Virginia, Charville, VA

solution in a certain, fixed, thin zone (0.2~0.6 cm) of the surface soil. Ingram and Woolhiser (1980) used a mixing-zone concept to model the chemical transfer into overland flow. The mixing depth was assumed to be a zone in which the rain, surface flow and chemical are intermixed. The release of chemicals was affected by rainfall intensity, soil surface slope, and rainfall kinetic energy. Ahuja et al. (1981) conducted runoff studies using soil boxes with permeable bottoms and concluded that there did not exist a zone of uniform and complete mixing on the surface soil layer. The degree of mixing they observed was greatest at the soil surface and decreased exponentially with depth. They proposed using an effective depth of interaction (EDI), within which the degree of mixing was uniform and equal to that at the soil surface.

Ahuja and Ross (1983), Lehman and Ahuja (1985) indicated that subsurface flow or interflow was one of the major factors to affect the solute transport. In their studies, the hydraulic conductivity of soil and surface slope control the interflow pathways. Parr et al. (1987) presented a diffusion-based model describing the release of chemical to runofff water without infiltration. They concluded that the mass transfer depended on the near-boundary interaction between overland shear flow and the soil interface. Richardson and Parr (1988) used this model and different sizes of uniform glass beads to experiment. The diffusion coefficient was found related to the shear Peclet number. These mass transfer phenomena were no more the molecular diffusion only. Wallach et al. (1988, 1990) used the diffusion-based soil model and mass transfer coefficient of laminar boundary layer to fit the experimental data reported by Ahuja and Lehman (1983). Good agreement was obtained between modelled and measured outflow concentration for relatively low infiltration rate. Lai et al. (1994) chose uniform glass beads and used a scouring-based model to simulate the transfer phenomena. The losses of chemicals were affected by Reynolds number of overland flow and the surface relative roughness.

In this paper, both two-stage scouring-based model and diffusion-based model were discussed and investigated by multiple regression of experimental data from a laboratory flume. Two different soil conditions were selected for testing in the same laboratory flume. The outflow concentration of soil experiments was compared with experimental data and results simulated by both two-stage scouring-based model and diffusion-based model.

The Two-stage Scouring-Based Model

A conceptual profile diagram of medium depth and surface runoff is shown in Figure 1. The media were initially saturated with solute of uniform concentration (C_0). When a constant flow rate of runoff passed over the medium surface,

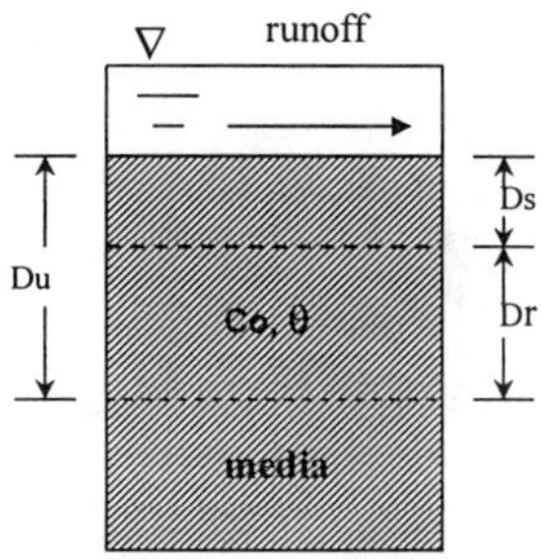

Fig. 1. The Conceptual Profile Diagram of Medium Depth and Surface Runoff

some medium water within thin layer near the surface would be scoured into surface runoff. The scoured layer depth (*Ds*) is a time-dependent variable, and can be expressed as:

$$Ds_i = Du_i - Dr_i \tag{1}$$

$$dDr_i / dt = -k_i Dr_i \tag{2}$$

where, *Ds* is the scoured layer depth, *Du* is the ultimate scoured depth, *Dr* is the residual scourable depth, *k* is the rate of change, *i* = 1 and 2 standing for the first and second stage, respectively, and *t* is time. Eq. (2) reperesents a first order decreasing of residual scourable depth. From (2), Dr_i can be solved as:

$$Dr_i = Du_i \ exp(-k_i \ t) \tag{3}$$

Assuming solute from the scoured layer would transport to downstream immediately, then, mass loss rate of chemical can be expressed as:

$$MLS_i = k_i Dr_i C_0 \theta \tag{4}$$

where *MLS* is mass loss rate, C_0 is the concentration of solute in residual scourable depth, θ is the porosity of media. Substituting (3) into (4),

$$MLS_i = (k_i Du_i C_0 \theta) \ \ exp(-k_i \ t) \tag{5}$$

let

$$IMLS_i = (k_i Du_i C_0 \theta) \tag{6}$$

Thus, (5) becomes

$$MLS_i = IMLS_i \ \exp(-k_i \ t) \tag{7}$$

k_i and $IMLS_i$ can be obtained from the diagram of semi-log mass loss rate of chemicals versus time, and then Du_i can be found from (6). The total head loss ($\Delta\Phi_{B\text{-}E}$) between the beginning and the end of medium packed bed can be calculated by Bernoulli's equation. The specific discharge (q) of subsurface flow within media can be estimated by Darcy's law:

$$q = K \Delta\Phi_{B-E} / L \tag{8}$$

where K is the hydraulic conductivity, and L is the length of packed bed. The hydraulic conductivity can be expressed as:

$$K = pg / v \tag{9}$$

where p is the permeability of medium, and v is the kinematic viscosity.

Let the specific discharge, q, represent hydraulic characteristic, L represent the characteristic length of packed bed, the kinematic viscosity (v) and fluid density (ρ_w) represent the characteristic of fluid, and the solute density (ρ) represent the solute characteristic. There are many values which can be chosen for the characteristic length of medium. Ward (1964) suggested that the square root of p can be used to represent the characteristics length of medium. Thus, the ultimate scoured depths (Du) and rates of change of scoured depth (k) of two stages can be expressed as two functions of these quantities:

$$Du_i = F1_i\left(q, \sqrt{p}, L, v, \rho, \rho_w\right) \tag{10}$$

$$k_i = F2_i\left(q, \sqrt{p}, L, v, \rho, \rho_w\right) \tag{11}$$

The Diffusion-Based Model

The conceptual diagram of diffusion-based model is shown in Figure 2. The media were also initially saturated with solute of uniform concentration (C_0). When a constant flow rate of runoff passed over the medium surface, the solute transfer through media follows the Fick's law.

$$\frac{\partial C}{\partial t} = D \frac{\partial^2 C}{\partial z^2} \tag{12}$$

where C is the solute concentration in aqueous solution at depth z, D is the diffusion coefficient. Because the surface runoff causes some local turbulence near the soil interface, the mass transfer is not the molecular diffusion only.

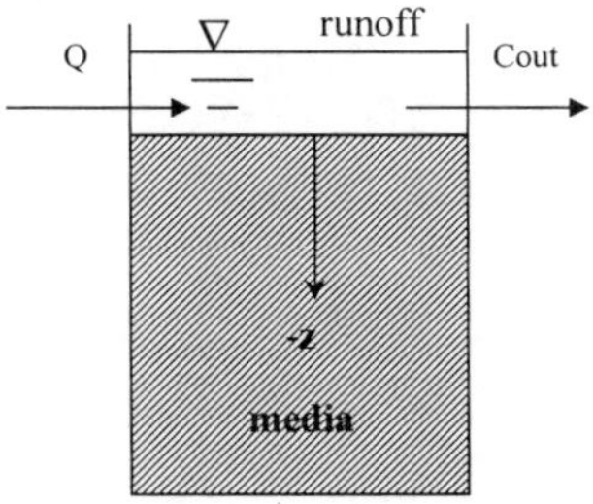

Fig. 2. The Conceptual Diagram of Diffusion-Based Model

At the interface (z = 0), the flux boundary and a entrainment coefficient (E) can be used to express the mass flux:

$$J(0,t) = -D(\partial C/\partial z)|_0 = E[C(0,t) - C_w(t)] \qquad (13)$$

where $J(0, t)$ is the mass flux at z = 0, C_w is the average solute concentration in surface runoff. Because C_w is too small compared with $C(0, t)$, C_w may be neglected (Ahuja & Lehman, 1983; snyder & Woolhiser, 1985):

$$J(0,t) = -D(\partial C/\partial z)|_0 = EC(0, t) \qquad (14)$$

Other boundary and initial conditions are

$$C(-\infty, t) = C_0 \quad , \qquad z = -\infty \qquad (15)$$

$$C(z, 0) = C_0 \qquad (16)$$

From (12), (14), (15) and (16), the solution can be obtained

$$C(z,t) = C_0[\exp(E^2t/D - Ez/D)\, erfc(\sqrt{E^2t/D} - z/2\sqrt{Dt}) - erf(z/2\sqrt{Dt})] \qquad (17)$$

The mass flux at z = 0, i.e. the source of solute to surface runoff, can be calculated:

$$J(0,t) = EC(0,t) = EC_0 \exp(E^2t/D)\; erfc(\sqrt{E^2t/D}) \qquad (18)$$

The water body of surface runoff can be assumed as a complete mixing zone. The mass balance can be expressed as

$$V\frac{dC_{out}}{dt} = A\theta J(0,t) - QC_{out} \tag{19}$$

where, V is the volume of surface runoff (= AH), A is the surface area of laboratory flume, H is the average water depth, C_{out} is the solute concentration of effluent, θ is porosity of soil, and Q is the flowrate of surface runoff. Assuming the initial concentration of effluent is zero. $C_{out}(0) = 0$, then

$$C_{out}(T) = C_0\xi\ [\exp(wT)\ erfc\sqrt{wT} - \exp(-T) + 2\sqrt{w/\pi}\Phi(\sqrt{T})]/(1+w) \tag{20}$$

where,

$$T = Qt/V \tag{21}$$

$$\xi = A\theta E/Q \tag{22}$$

$$w = (V/Q)/(D/E^2) \tag{23}$$

$$\Phi(x) = \exp(-x^2)\int_0^x \exp(y^2)dy \tag{24}$$

Laboratory Experiments

A series of experiments were performed in a horizontal, 370 cm long by 16.7 cm wide (W), laboratory flume (Fig. 3) without infiltration and non-rainfall conditions. Overland flow was passed at various velocities and depths over a 100 cm long (L) by 15 cm deep bed packed with uniform glass beads (Table 1) or soils (Table 2). Soil samples were taken from the experimental farm of National Taiwan University. All soils were air-dried at room temperature for three weeks. Stones and roots were removed and the remaining material was separated into two parts through 9.5 mm screen and 4.0 mm screen, and stored in labelled polyethylene bags to await analysis. Some physical properties of these two soils (G and F) are shown in Table 2. The permeabilities of glass beads and soils were measured by the column tests with several constant water heads.

Table 1. Characteristics of Glass Bead Media

Medium	Size (mm)	d_{50} (mm)	Porosity θ	Permeability (m^2)
A	2.90-3.50	3.25	0.38	1.86E-09
B	1.90-2.10	2.00	0.38	1.29E-09
C	0.95-1.20	1.05	0.37	6.10E-10
D	0.45-0.60	0.55	0.36	2.31E-10

Table 2. Physical Properties of Two Soils

Property	G (<9.5mm)	F (<4.0mm)
True density (g/cm^3)	2.65	2.65
Bulk density (g/cm^3)	1.10	1.14
Water content (%)	2.6	2.6
Porsoity	0.57	0.56
Effective porsoity	0.53	0.53
Permeability (m^2)	5.3×10^{-11}	1.4×10^{-11}

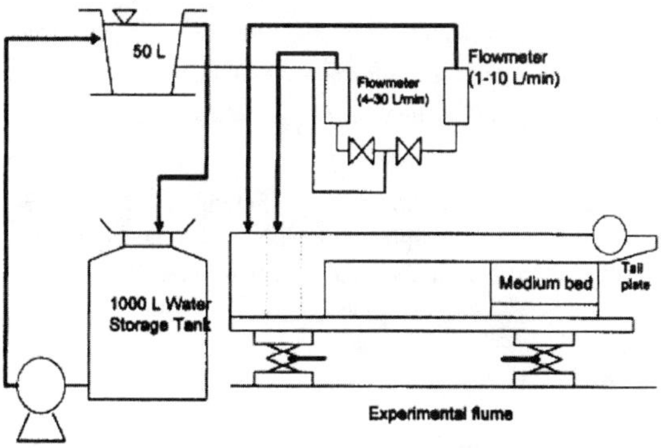

Fig. 3. Schematic of Experimental Setup

A constant head tank was used to supply water to the flume. The flow rate was controlled by a gate valve and a manometer column. This system can provide a stable flow rate throughout the whole experiment. The last 40 cm section directed the flow over an adjustable tail plate which can control the depth of runoff. A movable depth gage was mounted across the medium packed bed to measure runoff depth of three sections (the beginning, the middle, and the end) of the test part.

Before each run, the test section was saturated with 1,000 ppm, 5,000 ppm and 20,000 ppm potassium chloride solution for experiments of glass beads, and 10,000 ppm potassium bromide solution for soil experiments. A thin layer of solution above the surface was deemed satisfactory for these experiments. This thin layer of solution was flushed with flow initiation. Runoff samples were taken from the effluent weir overflow at every minute in first 10 min, every 2.5 min after 10 min to 30 min, and every 5 min after 30 min to the end of run typically. The Suntex SC-17A conductivity meter was used to analyze the concentration of potassium chloride or potassium bromide. Because the concentration of inflow or background concentration was stable, only four samples were taken from inflow in each run and analyzed.

Results and Discussions

Calibration of Two-Stage Scouring-Based Model

From experimental results, the first 10-min experimental data were calibrated to determine k_1 and Du_1, the other data were calibrated to determine k_2 and Du_2. The modelled results of Du_i and k_i of 88 runs were obtained. Generally, the decrease of mass loss rate of the first stage is faster than that of the second stage for each run, but the rates of change of two stages would be closed at high flow velocity. The ultimate scoured depth of the first stage is thinner than that of the second stage because the slope of the first stage is steeper than that of the second stage.

Nondimensionalizing all variables in (10) and (11), seven dimensionless parameters can be found, which are the dimensionless ultimate scoured depth ($Du_i^* = Du_i / \sqrt{p}$; $i = 1,2$), the dimensionless rate of change of scoured depth ($k_i^* = k_i p / v; i = 1,2$), the Reynolds number within media ($\boldsymbol{R}_p = q\sqrt{p}/v$), the relative length ($L^* = \sqrt{p}/L$), and the relative density ($\rho^* = (\rho - \rho_w)/\rho_w$). Only the determination coefficient (r^2) of each run over 0.8 (72 runs) was selected and analyzed with multiple regression to decide the relationships between the dimensionless modelled parameters (Du_i^* and k_i^*) and controlled variables ($\boldsymbol{R}_p$, L^* and ρ^*). Thus, Du_i^* and k_i^* can be expressed respectively by $\boldsymbol{R}_p$, L^* and ρ^* as:

$$Du_1^* = 10^{-2.4} \boldsymbol{R}_p^{0.60} (L^*)^{-1.5} (\rho^*)^{-0.16} \qquad (r^* = 0.62) \qquad (25)$$

$$k_1^* = 10^{4.1} \boldsymbol{R}_p^{-0.25} (L^*)^{2.4} (\rho^*)^{0.074} \qquad (r^2 = 0.71) \qquad (26)$$

$$Du_2^* = 10^{-4.2} \boldsymbol{R}_p^{0.69} (L^*)^{-2.0} (\rho^*)^{-0.31} \qquad (r^2 = 0.71) \qquad (27)$$

$$k_2^* = 10^{3.8} (L^*)^{2.2} (\rho^*)^{0.15} \qquad (r^2 = 0.79) \qquad (28)$$

The comparison of measured and calculated results from (25) to (28) are shown in Figs. 4~7, respectively.

Calibration of Diffusion-Based Model

The trial and error method was used to determine both entrainment and diffusion coefficients. Nondimensionalizing all results of D and E,

$$D^* = D / D_0 \qquad (29)$$

$$E^* = E\sqrt{p} / D \qquad (30)$$

where D_0 is the molecular diffusion coefficient of chloride (2.03 x 10^{-5} cm^2/sec). Analyzing all dimensionless parameters of D^*, E^*, $\boldsymbol{R}_p$, L^* and ρ^* by multiple regression, the relationships can be obtained:

$$E^* = 10^{10.1}(\boldsymbol{R}_p)^{-1.23}(L^*)^{3.45}(\rho^*)^{0.67} \qquad (r^2 = 0.66) \qquad (31)$$

$$D^* = 10^{-1.52}(\boldsymbol{R}_p)^{1.34}(L^*)^{-1.57}(\rho^*)^{-0.58} \qquad (r^2 = 0.76) \qquad (32)$$

The comparison of measured and calculated results from (31) and (32) are shown in Figs. 8, 9, respectively.

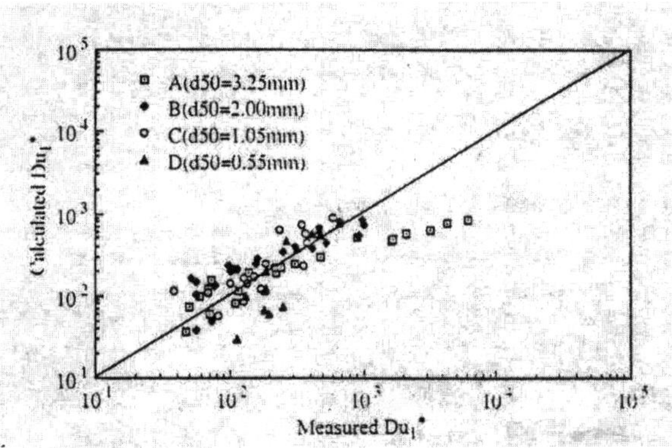

Fig. 4. Comparison of Measured and Calculated Du_1^* from Scouring-Based Model

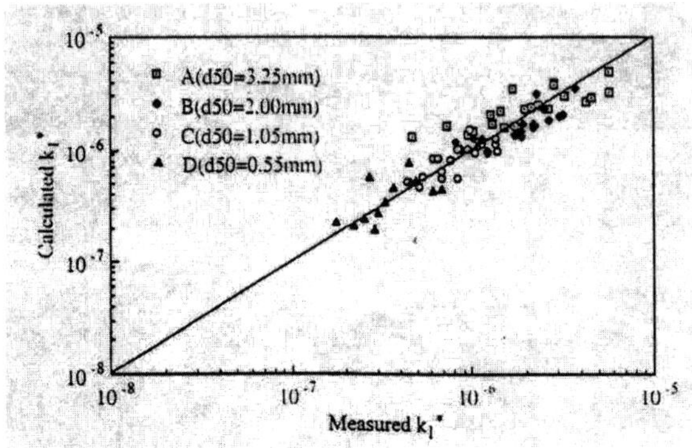

Fig. 5. Comparison of Measured and Calculated k_1^* from Scouring-Based Model

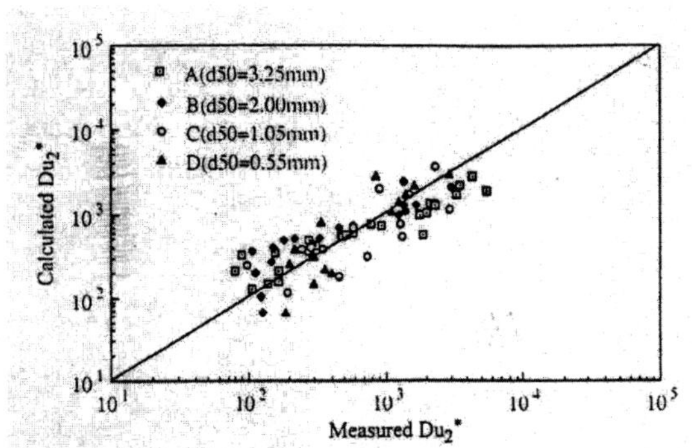

Fig. 6. Comparison of Measured and Calculated Du_2^* from Scouring-Based Model

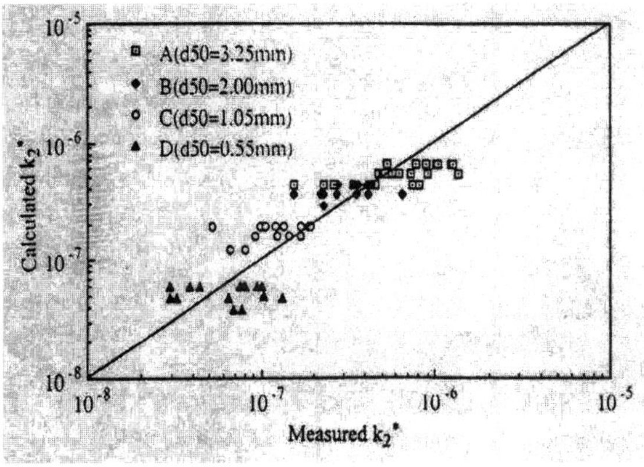

Fig. 7. Comparison of Measured and Calculated k_2^* from Scouring-Based Model

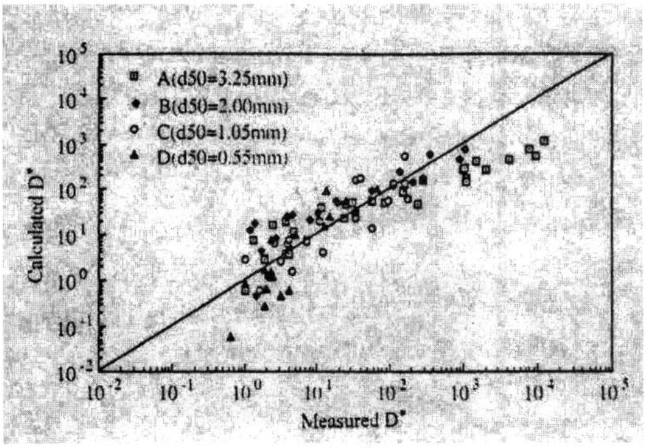

Fig. 8. Comparison of Measured and Calculated D^* from Diffusion-Based Model

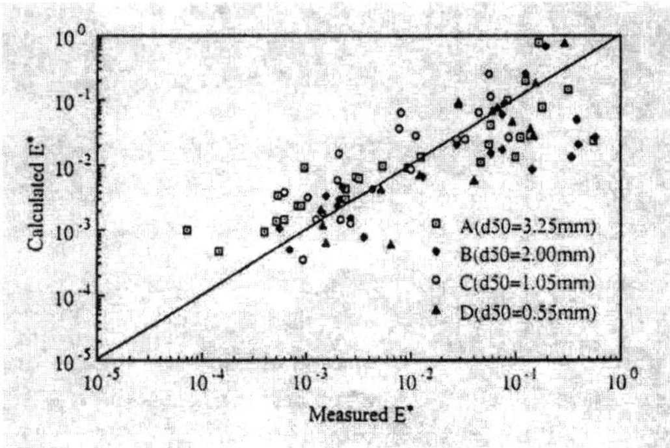

Fig. 9. Comparison of Measured and Calculated E^* from Diffusion-Based Model

Application of Two Models on Soils

From all physical properties and experimental hydraulic parameters of two soil conditions, the dimensionless ultimate scoured depth (Du_i^*) and the dimensionless rate of change of scoured depth (k_i^*) can be calculated from (25)~(28) by using the two-stage scouring-based model. All results of Du_i and k_i of two stages for each run are shown in Table 3. Substituting Du_i and k_i into the two-stage scouring-based model, the outflow concentrations of solute (bromide) can be obtained. Comparisons of modelled results and experimental results for each run are shown in Figs. 10 to 13.

Table 3. Calculating Parameters from Two-Stage Scouring-Based Model

Run No.	First stage		Second stage	
	Ultimated scoured depth Du_1 (cm)	Rate of change k_1 (min^{-1})	Ultimated scoured depth Du_2 (cm)	Rate of change k_2 (min^{-1})
G1	3.07E-01	7.86E-02	1.41E+00	1.21E-02
G2	6.11E-01	5.95E-02	3.03E+00	1.22E-02
G3	1.47E-01	1.08E-01	6.22E-01	1.23E-02
G4	5.14E-01	6.48E-02	2.50E+00	1.24E-02
F1	5.41E-02	1.44E-01	2.74E-01	1.07E-02
F2	1.83E-01	8.45E-02	1.06E+00	1.04E-02
F3	3.70E-02	1.70E-01	1.80E-01	1.07E-02
F4	9.98E-02	1.13E-01	5.40E-01	1.08E-02

From Table 3, all utlimate scoured depth of the first stage are less than 1 cm, agreeing with the mixing layer depth obtained from experimental data of glass beads. But some ultimate scoured depths of the second stage are larger than 1~3 cm. Thus, the agricultural surface runoff must consider all soil and hydraulic conditions to get more accurate results for long duration period. Comparing modelled results with experimental data, the simulation results of larger soil condition were good (Figs. 10 and 11), but the modelled outflow concentrations of finer soil condition were underestimated, especially for the initial period of runoff (Figs. 12 and 13).

Similarly, substituting all related properties of soils into (31) and (32), the diffusion coefficient (*D*) and entrainment coefficient (*E*) of the diffusion-based model can be obtained (Table 4). The molecular diffusion coefficient of Br^- is 7.2 x 10^{-4} cm^2/min from Bennett and Myers (1974). From (20), the outflow solute concentrations can be calculated. Comparisons of modelled results and experimental data for each run are shown in Figs. 14 to 17. All simulating results were underestimated, especially for the finer soil condition and initial period of runoff.

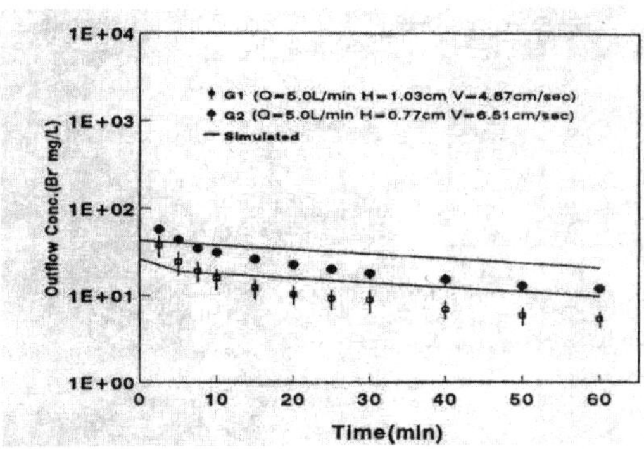

Fig. 10. Comparisons of Two-Stage Scouring-Based Model and Experimental Data for G_1 & G_2

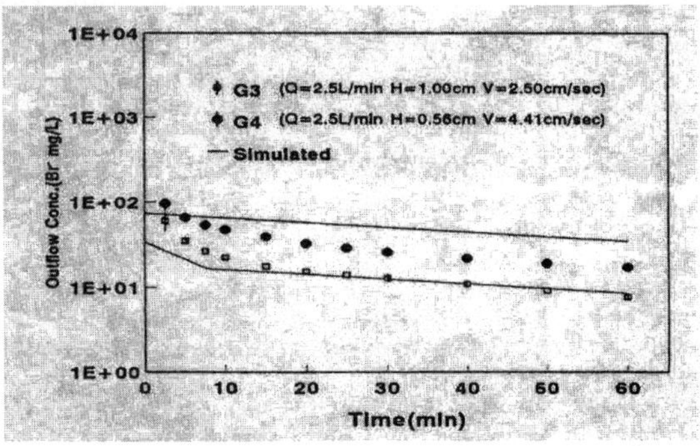

Fig. 11. Comparisons of Two-Stage Scouring-Based Model and Experimental Data for G_3 & G_4

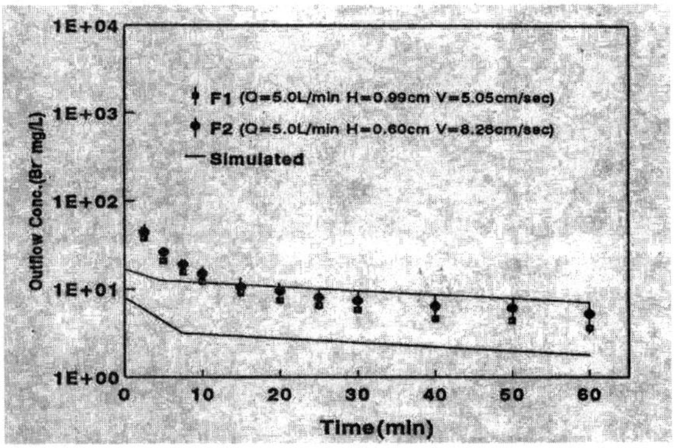

Fig. 12. Comparisons of Two-Stage Scouring-Based Model and Experimental Data for F_1 & F_2

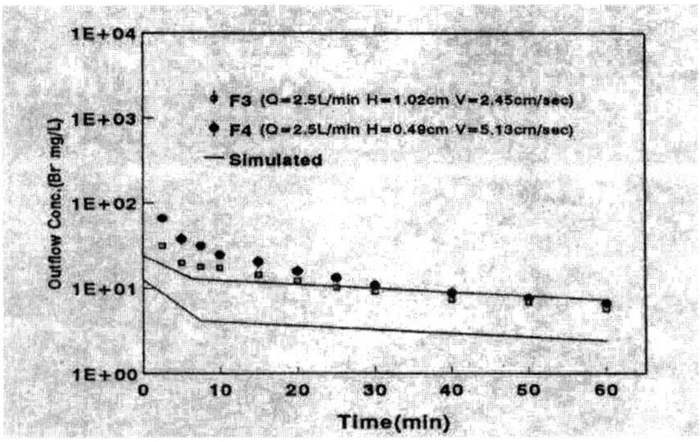

Fig. 13. Comparisons of Two-Stage Scouring-Based Model and Experimental Data for F_3 & F_4

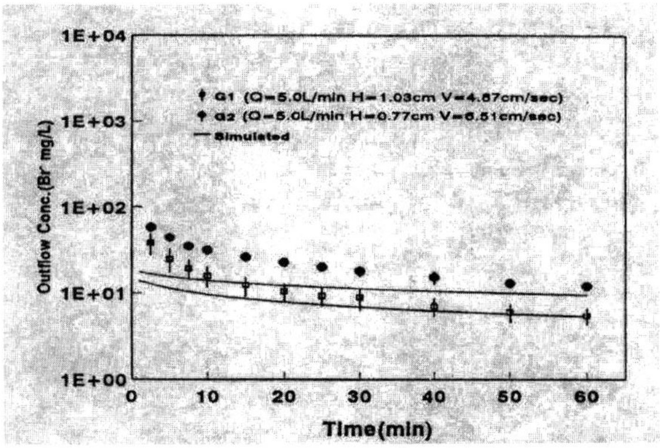

Fig. 14. Comparisons of Diffusion-Based Model and Experimental Date for G_1 & G_2

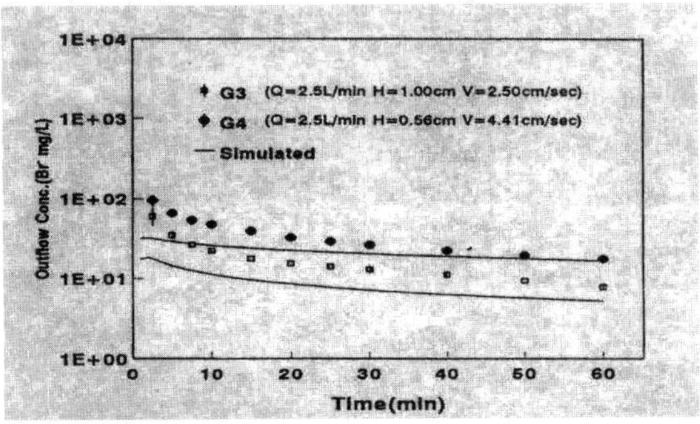

Fig. 15. Comparisons of Diffusion-Based Model and Experimental Date for *G3* & *G4*

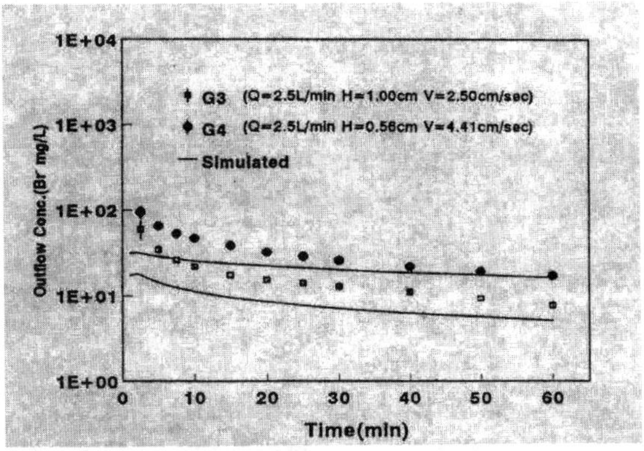

Fig. 16. Comparisons of Diffusion-Based Model and Experimental Date for F_1 & F_2

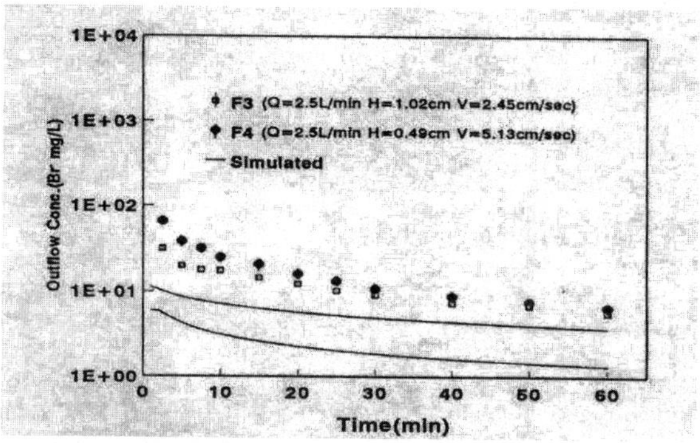

Fig. 17. Comparisons of Diffusion-Based Model and Experimental Date for F_3 & F_4

Table 4. Calculating Parameters from Diffusion-Based Model

Run No.	Diffusion coeff. D (cm^2/min)	Entrainment coeff. E (cm/min)
G1	4.97E-03	1.48E-02
G2	2.29E-02	1.68E-02
G3	9.64E-04	1.30E-02
G4	1.56E-02	1.63E-02
F1	1.40E-04	5.65E-03
F2	2.0E-03	7.05E-03
F3	6.04E-05	5.27E-03
F4	5.46E-04	6.31E-03

Summary and Conclusions

Both the two-stage scouring-based model and the diffusion-based model were presented and discussed to describe solute transfer to surface runoff. By use of the dimensionless analysis and multiple regression of experimental data from a flume packed with glass beads, dimensionless Du_i^* and k_i^* for the two-stage scouring-based model, and D^* and E^* for the diffusion-based model were obtained by expressing as functions of $\boldsymbol{R}_p$, L^* and ρ^*. Replacing the glass beads with soils, modelled results and experimental data were compared. Since some ultimate scoured depths of the second stage are larger than 1-3 cm, i.e., larger than the mixing layer depth reported by some researchers using the mixing-zone concept, the agricultural surface runoff must consider all soil and hydraulic conditions to get more accurate results for long duration period. Using the two-stage scouring-based model, simulation results of larger soil condition were good, but the modelled outflow concentrations of finer soil condition were underestimated. For the diffusion-based model, all simulating results were underestimated.

Acknowledgement

The authors would like to experess their thanks for the financial support provided by National Science Council, Taiwan, ROC.

References

Ahuja, L. R,, Sharpley, A. N., Yamamoto, M., and Menzel, R. G. (1981). "The depth of rainfall-runoff-soil interaction as determined by ^{32}P." *Water Resour. Res.,* 17, 969-974.

Ahuja, L. R., and Lehman, O. R. (1983). "The extent and nature of rainfall-soil interaction in the release of soluble chemicals to runoff." *J. Envir. Quality,* 12(1), 34-40.

Ahuja, L. R., and Ross, J. D. (1983). "Effect of subsoil conductivity and thickness on interflow pathways, rates and source areas for chemicals in a sloping layered soil with seepage face." *J. of Hydrology,* 64, 189-204.

Bennett, C. O., and Myers, J. E. (1974). *Momentum, Heat and Mass Transfer.* McGraw-Hill, New York.

Donigian, A. S., Beyerlein, D. C., Davis, H. H., and Crowford, N. H. (1977). Agricultural runoff management (ARM) model version II: refinement and testing. Environmental Research Laboratory, Athens, Georgia, EPA 600/3-77-098.

Haith, D. A. (1986). "Simulated regional variations in pesticide runoff." *J. Envir. Quality,* 15(1), 5-8.

Ingram, J. J., and Woolhiser, D. A. (1980). "Chemical transfer into overland flow." In: Pro. of Symposium on Watershed Management, Boise, ID, July 21-23, ASCE, New York, 40-53.

Lai, J. L., Lo, S. L., and Lin, C. F. (1994). "Effects of hydraulic and medium characteristics on solute transfer to surface runoff." *Wat. Sci. Tech.,* 30(7), 145-155.

Lehman, O. R., and Ahuja, L. R. (1985). "Interflow of water and tracer chemical on sloping field plots with exposed seepage faces." *J. of Hydrology,* 76, 307-317.

Li, W., Merrill, D. E., and Haith, D. A. (1990). "Loading function for pesticide runoff." *J. Water Poll. Control Fed.,* 62(1), 16-26.

Miller, M. H., Robinson, J. B., Coote, D. R., Spires, A. C., and Draepr, D. W. (1982). "Agricultural and water quality in the Canadian Great Lake basin: III. phosphorus." *J. Envir. Quality,* 11(3), 487-493.

Owens, L. B., Edwards, W. M., and Vankeuren, R. W. (1984). "Peak nitrate-nitrogen values in surface runoff from fertilized pastures." *J. Envir. Quality,* 13(2), 310-315.

Parr, A. D., Richardson, C. P., Lane, D. D., and Baughman, D. (1987). "Pore water uptake by agricultural runoff." *J. Envir. Engrg., ASCE,* 113, 49-63.

Richardson, C. P., and Parr, A. D. (1988). "Modified Fickian model for solute uptake by runoff." *J. Envir. Engrg., ASCE,* 114, 792-809.

Sharpley, A. N., Smith, S. J., Berg, W. A., and Williams, J. R. (1985). "Nutrient runoff losses as predicted by annual and monthly soil sampling." *J. Envir. Quality,* 14(3), 354-360.

Snyder, J. K., and Woolhiser, D. A. (1985). "Effects of infiltration on chemical transport into overland flow." *Trans. ASAE,* 28, 1450-1457.

Ward, J. C. (1964), "Turbulent flow in porous media." *Hydraulics Divison, ASCE, Hy* 5, 1-12.

Wallach, R., Jury, W. A., and Spencer, W. F. (1988). "Transfer of chemicals from soil solution to surface runoff: a diffusion-based soil model." *Soil Sci. Soc. Am. J.,* 52, 612-618.

Wallach, R., and Genuchten, M. T. (1990). "A physically based model for predicting solute transfer from soil solution to rainfall-induced runoff water." *Water Resour. Res.,* 26, 2119-2126.

AUTHOR INDEX